Introduction to Copper Cabling

Application for telecommunications, data communications and networking

Introduction to Copper Cabling

Application for telecommunications, data communications and networking

John Crisp

Newnes

OXFORD   AMSTERDAM   BOSTON   LONDON   NEW YORK   PARIS
SAN DIEGO   SAN FRANCISCO   SINGAPORE   SYDNEY   TOKYO

Newnes
An imprint of Elsevier Science
Linacre House, Jordan Hill, Oxford OX2 8DP
225 Wildwood Avenue, Woburn, MA 01801-2041

First published 2002
Copyright © 2002, John Crisp. All rights reserved

British Library Cataloguing in Publication Data
A catalogue record for this book is available from the British Library

Library of Congress Cataloging in Publication Data
A catalog record for this book is available from the Library of Congress

ISBN 0 7506 5555 0

For information on all Newnes publications visit our
website at www.newnespress.com

Typeset by Keyword Typesetting Services Ltd
Printed and bound in Great Britain by Biddles Ltd, *www.biddles.co.uk*

Contents

Preface

How will future generations refer to our times? Will it be known as one of space exploration, genetics, atomic energy or computing? Possibly, but I think it is more likely to be 'The age of communications'. Not since printed books and newspapers were first introduced has there been such an explosion of communication. None of this technology could function without modern cables and, just as important, competent installers.

In every building, there is a lot of technology hidden above the ceiling, under the floor and in the walls, as a network of cables interconnects every desk, room and beyond, across the city and the world.

As with learning all forms of new technology, we have the problem of getting started. We need to know the basic nuts and bolts of how it works and face the problem of 'the words'. There always seems to be hundreds of new words that everyone else seems to understand.

This book introduces cabling without assuming any previous knowledge of the basics or the jargon. I hope you find it both useful and an enjoyable start to your life in cables.

John Crisp

1

Talking across the Atlantic

Unless we were to remain happy to communicate by lighting bonfires on hilltops or employing runners or galloping horses, we needed a real breakthrough.

A good idea that wasn't followed up was the work of the Greek Thales who, in the sixth century BC, was one of the first to investigate electricity and magnetism. He did the trick of rubbing a balloon on his sleeve, and picking up some light pieces of material – actually he rubbed a piece of amber with some fur, but that's near enough the same.

If he could only have encouraged more people to continue his work, we may have had warp drives and teleporters by now. But no-one showed any interest and, like they say in all good history books, 'nothing happened for 2300 years'.

In the sixteenth century there was a flurry of interest in devising some form of long-distance communication system. There were many attempts with lights and mirrors, which had limitations where distances were involved. But flashing lights were nowhere as limited as the speaking tube, in which we would shout down a tube with an ear placed at the other end. This would limit the scope for long-distance communications but was resurrected about 200 years later with the idea of banging the outside of the tube with a hammer to send messages in code. This didn't work either.

Towards the end of the sixteenth century and in the early seventeenth century, we had noticed that the effects of magnetism and many people had great faith in its ability to provide a fast long-distance communication system as well as miracle cures for all ills and almost anything else.

It had been noticed that a compass needle can be deflected by another compass needle and so it was a small step to conclude that if it could be made to happen at any distance then all communication problems are virtually solved. There was a tendency to assume that this small problem would soon be surmounted and we could get on with the real invention bit. Rumors and claims were made for the most unlikely methods.

One very popular method was to make two such needles become 'sympathetic' and by moving one, the other would move by the same amount to remain parallel even when separated by enormous distances. Having produced two sympathetic needles, all we have to do is to mark the edge of the compass with the alphabet, and there we are – no-cost instant communications. Despite the small fact that the needles never did align themselves in parallel even when the needles were close together, the rumor spread quickly, probably just because it was such a neat idea that everyone wanted it to be true. A bit like flying saucers and little green men from Mars.

In those days we had to spread rumors from person to person but nowadays we have the media, which are in the fortunate position of not only being able to spread rumors to millions of people at a time but actually get paid for doing so.

When it was accepted that the range was very limited and the needles never did remain parallel, we could still demonstrate that a magnet could be used instead of the first 'transmitting' needle and the receiving needle would still be deflected, so the rest of the system could still work. It was soon seen to be limited in range but, and this gave hope, a larger magnet would have a larger range. All we needed to do was to build larger and larger magnets but this idea also died as some calculations were done on the size of the magnet needed for transmission over a few hundred miles. There was also the small problem of interference if more than one communication system was set up in the same area.

The American scientist, Benjamin Franklin was rumored to investigate electricity by flying a kite in a thunderstorm. Now, whether he actually did this or not, it was certainly a good enough story to encourage others to try it and to die in the attempt.

Benjamin certainly did investigate and develop the idea of air terminal or lightning conductors on buildings but even nowadays, many people have the wrong idea of what they are for and how they work. We will look at lightning conductors in a later chapter.

About 50 years later, in Geneva, Switzerland, George Lesage decided to use wire and a pith ball. The point of the pith ball was that it was extremely light and when charged with electricity by a connecting wire they would hold equal polarity of charge and hence repel each other. The pith ball would move away from the wire as in Figure 1.1. If we connect one wire for each letter of the alphabet, all we had to do was to energize the wire corresponding to the first letter and, at the far end, they would watch to see which ball moved. By this laborious process we could spell words, one letter at a time.

Figure 1.1
Really slow telegraphy – but it worked

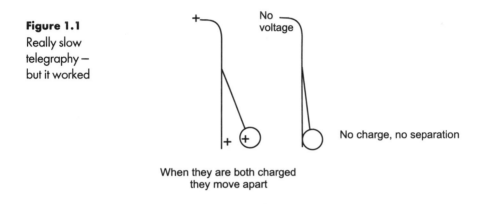

No voltage

When they are both charged they move apart

No charge, no separation

Electricity is too slow – let's try something mechanical

Like most methods up to this time, the search was always for a way to generate movement at a distance, usually to point to a letter. This latest version abandoned electricity in favor of mechanical engineering. As usual, it was a good idea on paper but more difficult to actually achieve. This is how it worked, or should have worked. To send a message to your house, all we have to do is to pop down to my cellar, turn a lever until it pointed to the first letter and a system of mechanical gears and drive shafts between our two houses turned the pointer in your cellar. Then on to the next letter. It would probably work if we lived in adjacent houses but there would be rather a lot of friction if we were in different towns. Not to mention the small problems of hills and rivers.

Five years later in 1792 another mechanical solution by Claude and Rene Chappe took a step back towards the ridiculous but within three years of further development they had a winner.

It worked, but the neighbors were not happy

Behind their parent's house, the two brothers arranged a clock face with just a single hand that swept around the face in 30 seconds. Ten numbers were written on the face so every three seconds the hand pointed to one of the numbers. At the receiving station a few hundred yards away, a similar clock face was set up. The first job was to synchronize the sweeping hand. This was done very simply by striking a casserole dish as the hand passed the vertical position, whereupon the hand on the receiving apparatus was released. A second clang indicated the moment to read the transmitted number.

To the irritation of all in earshot, a series of numbers could be clanged out and hence deciphered by reading the position of the receiving hand. By coding the numbers into letters, a message could be beaten out.

It was soon modified to a system that was purely visible, which must have been an enormous relief to all those within earshot. This was a wooden panel that was painted black on one side and white on the other and the moment that corresponded with the required number was signaled by pivoting the board to change the color.

The day of the big test came in March 1792. With the aid of a telescope, and in only four minutes a message, 'if you succeed you will soon bask in glory', was sent over a distance of 10 miles. The message was supplied at the moment of the test to avoid trickery which, as we can imagine, was rife with the demonstration of new signaling methods.

That's better

The Chappes were happy to see the system work over such distances but synchronizing the clock was a nuisance so they thought of ways to eliminate the timing problem. But was it possible? In 1793, the Chappe Mark III version was unveiled and this was a winner.

The clock was thrown out and instead a semaphore system was built on a tower. The tower had two pivoted arms on it and provided enough combinations to provide a different pattern for each letter and number. A system of ropes was used to reposition the arms as shown in Figure 1.2 and the towers were used to send a message through a series of such towers over 20 miles and to everyone's amazement, 20 minutes later, a reply was received. Not a bad speed.

Claud Chappe was put on a government salary to build a series of towers over hundreds of miles around France. They even built one on the French coast to enable easy communications with Britain after the forthcoming invasion. As it happened, they never did manage to invade, so the British end of the link was never constructed. During this

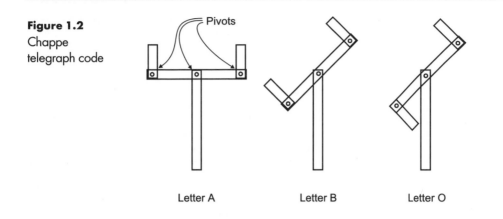

Figure 1.2
Chappe
telegraph code

Letter A   Letter B   Letter O

time the British were busy building their own set of towers of a slightly different design to connect the Naval Headquarters at London with the south coast ports. They were also busy trying to burn down the French towers.

During the next 20 years, over 1000 towers were built across Europe.

Yes, very nice, but it could be faster

For speed, ideas were returning towards electricity. We had the usual crop of interesting but impractical ideas. There was one in which the ends of 26 wires were dipped in acid and, when energized, bubbles would appear and the letter could be deciphered.

Francis Ronalds set up a working electric telegraph system around his garden which used a clock similar to the Chappe casserole lid method except that a burst of electricity moved a pith ball to indicate the letter being sent. He contacted the British Navy who refused to attend a demonstration saying 'telegraphs of any kind are now wholly unnecessary'. This could be a contender for the 'Quote of the Millennium'.

Would instantaneous transmission be fast enough?

About 70 years previous to this time, an experiment had been carried out to test the speed of electricity along a wire. In April 1746 Jean-Antoine Nollet, a French scientist, received the cooperation of about 200 monks for an electrical experiment. Now, whether they volunteered through interest, ignorance or under duress is not known but they were connected to each other by a 25 foot (about 7 metres) length of iron wire and formed a line over a mile in length. Monsieur Nollet gave them an electric shock and watched the effect – he saw that every monk appeared to be affected at the same time. Therefore electricity moves instantaneously around a circuit. We now know that this is an overestimate but it is certainly very fast.

Then came Morse and Cooke

Morse was a painter from Charlestown, Massachusetts and was looking for a way to make a fortune. He was on a six-week voyage from Europe when, by chance, he had a conversation about electricity with a fellow traveler. His companion just happened to mention that it was thought, thanks to the monks, that electricity traveled instantaneously along a wire. It occurred to Morse that if electricity could travel at that speed, it would be ideal for sending information.

Now, this idea was new to Morse and he had no idea that scientists had spent most of the last century grappling with just this problem. Assuming that the electrical problems would be easy to solve, he ignored them and got started on the code that could be used to send the information.

Four years later, in England, William Cooke had recently ended his career with the British army in India and was looking for a way to make his fortune. He had recently attended a lecture on electricity and was so struck by the possibilities that he started experimenting immediately.

Morse and Cooke worked independently but they both met the same problem. Regardless of how large a battery they used, they could only make the electricity flow along short lengths of wire and neither of them had enough scientific knowledge to make any progress.

As it happened, only one man at that time knew the answer. He was an American physicist called Joseph Henry who had solved the problem by connecting batteries in series to create a higher voltage, but neither Morse nor Cooke knew that.

We now find a strange coincidence. Cooke asked for help from Charles Wheatstone who not only knew about Henry's work but also just happened to have several miles of wire available for experiments. In America, Morse was asking around to try to find someone with the scientific knowledge to help. He first approached a chemist who just happened to be a personal friend of Henry – so Morse was given the solution as well.

A good idea was one thing, but selling it was another

Morse had some luck. In 1838, Congress was to build a Chappe-style telegraph system from New York to New Orleans and asked interested parties for their comments. He traveled to Washington complete with his demonstration, which by this time had succeeded over a distance of

10 miles. Unfortunately, Congress, with all the foresight of the British Navy, lost interest in the project.

Getting nowhere in the USA, Morse came to Britain and then to Europe and getting nowhere there either, he returned to the USA. Congress had a change of heart and financed a line from Washington to Baltimore. They still had reservations and appointed a John Kirk to keep an eye on Morse who, it was suspected, was a little crazy.

Kirk became converted to the idea of telegraphy and when the Whig National Convention was held in Baltimore, he arranged for details to be sent by telegraphy to Washington, and the crowd that gathered heard all the results over an hour before the arrival of the train.

Meanwhile, Cooke had begun to make some sales to railway companies but in a close parallel to the Baltimore situation, the telegraph enabled news of the birth of Queen Victoria's son to appear in *The Times* newspaper, on sale in the streets of London within 40 minutes of the announcement of the birth. Another moment of notoriety came when telegraphing the description of a wanted murderer allowed the police to be waiting as he stepped off the train. He probably became the world's first technophobe.

By 1850, there were over 2000 miles of wires in the USA and a similar amount in the UK. Other European countries were also investing in similar systems. The telegraph was now established.

Water and electricity don't mix

Around this time, Morse also built an underwater cable that was sheathed in rubber and protected by an outer lead pipe. He laid it between the Battery and Governor's Island in New York Harbor.

About the same time, an ambitious plan was being hatched to connect a cable from France to England, about 25 miles under the English Channel.

Rather than use the rubber and lead method like Morse, they decided to coat a wire with only ¼ inch (6 mm) of gutta-percha. This was a gum derived from a tree that had the useful property of a low melting point so hot water would allow it to be molded and then solidify in cold water. A sort of early moldable plastic.

With great optimism, a ship set off from England to France unwinding a drum of cable as it went. It was not long before another property of gutta-percha was discovered – it was very buoyant and weights had to be added to keep the cable on the seabed. After completing the journey, the link was complete and some messages were sent in celebration, though the rejoicing was short-lived.

The very next day, the cable company was introduced to a problem that was to worry it for years, as a fishing boat hooked up the cable and took a bit home out of curiosity. The replacement was very heavy and well reinforced with iron. In deep water it tended to run quickly off the drum and they ran out of cable before they reached France and had to splice on an extra length. Even so, the cable worked and in 1852 it was opened for business.

Installed cables had now reached thousands of miles on land and underwater ran from England to Ireland and crossed the Mediterranean Sea but people still stared out over the Atlantic Ocean and wondered.

The problems and the costs

Anyone who knew anything about telegraphy or science would know that the idea of sending electrical signals along a waterproof cable all the way from England to America was about as silly an idea as could be imagined. And the cost would be phenomenal. Only a rich and ignorant person would even dream of it.

And along came Cyrus Field, a person conveniently rich and quite ignorant of the problems of telegraphy – just what we needed. It happened that an English engineer, having recently failed in his attempt to build a line across Newfoundland, was looking for a backer to connect New York to Newfoundland.

Our Cyrus got so excited about this new telegraphy idea that they soon decided to go straight for the big one: New York to Newfoundland and straight across the Atlantic. After a couple of years, the first leg was done.

In England, they met up with Morse and a John Brett who together had a dummy run by connecting together 2000 miles of cable and, being able to send messages along it, were suitably encouraged.

What they needed now was a first-class engineer to design the real cable. What they got was the completely incompetent Edward Whitehouse. Considering the poor insulation available, it was not a good idea to use very high voltages to pass the information. Worse than this, the resistance was high due to the thin wires that had been specified.

The 2500 ton cable was carried in two separate ships but broke four times as it was laid. Eventually, to save time, two ships met in mid Atlantic and one headed for Ireland and the other for Newfoundland. The cable came ashore and, amid much jubilation, it failed. It took a further week to get it working again, and the inaugural message took the whole day to send. The cable finally died a month later.

The new cable

The disgraced Whitehouse was replaced by Professor Thomson, and a new and ultimately successful cable was designed. The largest ship in existence, the *Great Eastern*, set out from Ireland laying the new cable and a couple of weeks later successfully reached Newfoundland.

On the first day of operation it proved so popular that a new cable was planned immediately and was laid within a few months to double the transatlantic capacity.

A trend that continues today.

2

Technical bits that may be useful

At the end of many of the chapters, there is a small quiz to allow us to try out our new skills. All the answers are given at the back of the book with full workings for calculations. It's just a bit of fun – don't take it seriously.

A small charge

We are used to buying fuel by the gallon or liter, and newspapers and books are sold 'each' but what about electricity?

The smallest quantity of electricity is a single electron but it would be difficult to sell electricity by the electron just as it would not be practical to sell rice by the grain. The unit we use is about 6.25 million million million electrons and we call this a 'coulomb' of electricity. When we create static electricity by rubbing a balloon on our sleeve we are transferring electrons from our sleeve to the balloon, or the other way around. The absence of electrons also creates an electric charge. We know that the charge left on the balloon is equal and opposite to the charge on our sleeve and so it was convenient to call the two types of charge positive and negative. The electron was given the title of negative charge and the absence of electrons was called a positive charge.

We have all tried sticking magnets together and discovered that two similar poles will repel each other and opposite poles will attract. The same occurs with charges – unlike charges attract and like charges repel.

The current flows (direct current or DC)

When there is a negative charge and a separate positive charge, given the chance, the electrons will flow towards the positive charge until the positive and negative charges cancel each other. This is very much like water flowing through a pipe from a high water level to a lower level, as shown in Figure 2.1. The water will flow until the two levels even out and then everything stops.

Figure 2.1
In some ways, water and electricity are similar

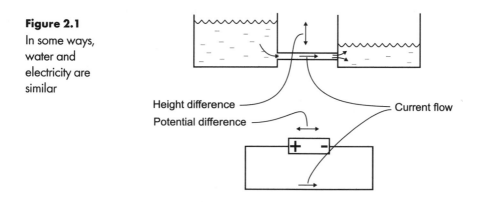

The larger the difference in water level, the more water that flows every second along the pipe. In the case of the pipe, we may measure the flow in gallons or liters per second, and in electricity it is measured in coulombs per second, which we call amps (A). The electrical equivalent of increasing the difference in the water levels is increasing the voltage (V), which is also called the potential difference (p.d.).

How much water or electrical current (I) that actually flows every second will also depend on how easy it is for the water or electrical current to flow. We can easily imagine that, all else being equal, a pipe of diameter of, say 1 inch (25 mm) would pass less water that one with a diameter of 1 ft (300 mm). Electricity is similar to water but not as nice to drink.

In an electrical circuit, the ease with which the electricity can flow depends on the dimensions of the cable and also on the material used. How difficult electricity finds it to move along the cable is called the resistance (R), measured in ohms (Ω). The longer and thinner a conductor is, the more resistance and hence the smaller the current that will flow.

The voltage which results from current flowing through a resistance is given by the relationship known as Ohm's law: volts in volts = current in amps × resistance in ohms, usually written as $V = IR$.

Conventional current and electron flow

When we switch on a light bulb, the current flows through the bulb – we reasoned this out before we knew about electrons and we had to guess which way the current was flowing. Unfortunately, we guessed wrong and assumed that, whatever electricity was, it flowed from positive to negative. We then found out that it was actually electrons moving from negative to positive but by this time, the positive to negative idea was so well established that it proved impossible to change – so we have stayed with it.

If we are talking about current flowing in a circuit, we refer to positive to negative current which we call 'conventional' current or just 'current'. However, if we get into itsy-bitsy details about what is happening inside a component we often use electron flow which is from negative to positive. If in doubt, always assume conventional current.

Alternating current (AC)

If we increase and decrease the voltage of a battery, the value of the current will increase and decrease, and also if we reverse the polarity of the voltage, the direction of the current will reverse.

All of the mains electricity supplies are generated by revolving machinery whether fueled by nuclear, wind, gas, coal or oil. The direct result of the design of the generators is that the output voltage varies between a peak positive and negative value in the form of a sine curve. This is shown in Figure 2.2.

Figure 2.2
A sinusoidal (or sine) wave

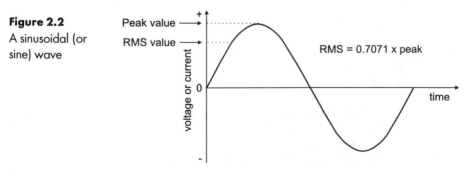

RMS voltage

If we applied a DC voltage to an electric heater and carefully measured its temperature, we could then disconnect the DC voltage and replace it by a small AC voltage. Measuring its temperature as we slowly increased the size of the AC voltage until the temperature reached the same value, we could say that the power supplied by the DC and AC voltages is the same.

This is called the 'RMS' value of the AC voltage. With a sine wave, it works out to 0.7071 × peak (maximum) value of the sine wave. This figure of 0.7071 only refers to a perfect sine wave.

To calculate the DC equivalent, or RMS value of any other shape waveform, we would have to do some serious maths. We would need to plot a graph showing the SQUARE of each value on the shape; then find the average or MEAN of all the readings. Finally, take the square ROOT to get the RMS (Root Mean Square) or DC equivalent value for the waveform. It's interesting to see where the term RMS comes from, but the good news is that in the last 20 years I have never had to do the calculation.

Peak voltage

If an AC voltage is just quoted as, say, 220 volts (220 V), we mean that it is 220 V RMS. If we know the RMS value, the peak or maximum voltage is higher than this – actually 1.414 × greater so a 220 V mains supply actually reaches 220 × 1.414 = 311 V.

In fact, it will vary from +311 V to −311 V.

Frequency (f)

The frequency is the number of complete cycles that the voltage goes through in a second. This applies equally well to sound waves or electrical signals.

Phase

This is the alternating voltage (or current) equivalent of marching in step. If two waveforms are in phase, they start at the same moment and stay together, as in Figure 2.3. Their amplitudes are not important.

Wavelength

If a signal is moving along a cable at, say, 200 million meters per second then a signal that is changing at the rate of one cycle per second, or 1 hertz (Hz), would have traveled 200 million meters during the time

Figure 2.3
Marching in
step

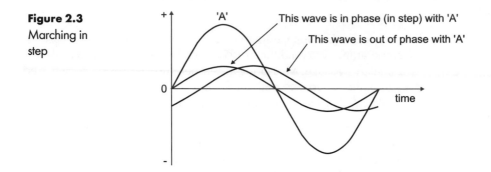

taken for the completion of the sine wave. This distance is called the wavelength of the signal. Wavelength is given the symbol λ (lambda).

The formula which connects the speed (s), wavelength (λ) and frequency (f) is $s = f\lambda$. By transposing, we can get the other variations: $f = s/\lambda$ and $\lambda = s/f$.

Multiples and sub-multiples

There are many terms but the ones that we will come across are:

giga (G)	one thousand million (a billion)	1×10^9
mega (M)	one million	1×10^6
kilo (k)	one thousand	1×10^3
milli (m)	one-thousandth	1×10^{-3}
micro (μ)	one-millionth	1×10^{-6}
nano (n)	one-thousandth of a millionth (a billionth)	1×10^{-9}
pico (p)	one-millionth of a millionth	1×10^{-12}

Frequency spectrum

The full range of frequencies is divided into bands, the use of which is internationally controlled.

The range of a transmission is determined much more by the frequency used than by the power. The exception is very low frequencies, which can obtain world-wide range, with sufficient power, as the waves tend to follow the curvature of the Earth.

Between about 4 and 30 MHz, long range transmission is the result of reflecting the signals between the Earth's surface and the ionized regions in the ionosphere which act like conducting mirrors. The ionized regions are maintained by energy from the Sun and so vary between day and night and summer and winter.

Above 30 MHz, no reflection takes place and the signals travel in a straight line and so the range is limited by the distance to the horizon. It is for this reason that VHF and UHF (very- and ultra-high frequency, respectively) transmitters are installed on high towers and buildings to provide maximum range. The bands are shown in Table 2.1.

Capacitance

Have a look at Figure 2.4. We have two wires connected to a battery but they are not touching – what would happen? Would any current flow?

When the switch closes, the chemical energy in the battery will force electrons from the negative terminal along the wire. As the electrons move along the wire, they repel the electrons in the other piece of wire. The effect of moving the electrons away from the wire on the right-hand side is to leave too few electrons and this will cause a positive charge. Just as we found with the water levels in the tanks, the current will stop flowing as soon as the voltage between the ends of the wires is equal to the battery voltage.

Now, if we disconnected the battery, the left-hand wire would have too many electrons trapped in it and the right-hand wire would have too few electrons in it, as shown in Figure 2.5.

The storage of charge between two conductors is called capacitance. Making a connection between the two wires will allow the excess elec-

Table 2.1 Frequency spectrum

Symbol	Name	Band	Frequency range	Transmission range
VLF	Very low frequency	Audio	3–30 kHz	World-wide. Power limited
LF	Low frequency	Audio	30–300 kHz	World-wide. Power limited
MF	Medium frequency	Radio (RF)	300 kHz–3 MHz	Typically 800 km (500 miles)
HF	High frequency	Radio (RF)	3–30 MHz	World-wide at best time of day/season
VHF	Very high frequency	Radio/television communications	30–300 MHz	Line of sight
UHF	Ultra high frequency	Radio/television communications	300 MHz–3 GHz	Line of sight
SHF	Super high frequency	Radar, satellites	3–30 GHz	Line of sight
EHF	Extra high frequency	Radar	30–300 GHz	Line of sight

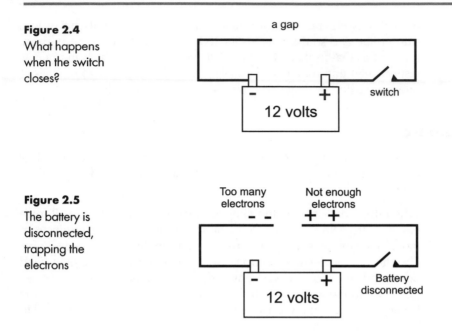

Figure 2.4
What happens
when the switch
closes?

a gap

- **+**
12 volts

switch

Figure 2.5
The battery is
disconnected,
trapping the
electrons

Too many
electrons
- -

Not enough
electrons
+ +

- **+**
12 volts

Battery
disconnected

trons to flow back around the circuit until the voltage difference between the two wires is reduced to zero so the capacitance between the ends of the wires will be acting just like a battery.

The energy is actually stored in an electric field that always connects two conductors at different potentials and, when drawn as in Figure 2.6, the arrows represent the direction of the field from positive to negative. Again, this is a throwback to conventional current.

Capacitance and capacitors

Capacitance occurs between any two places that are at different voltages and since the electric field stores electricity, we can build devices that store a known amount of energy. These devices are called capaci-

Figure 2.6
An electric field
is formed

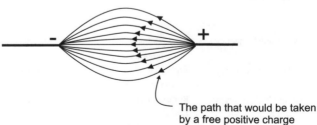

− +

The path that would be taken
by a free positive charge

tors. There are many different designs but a simple one would be made from two sheets of aluminum foil separated by a plastic sheet and rolled up to keep it small.

If we apply a continuously changing voltage like a data signal or an alternating voltage, the current flows into or out of the capacitance continuously, giving the impression that the signal can pass through a capacitance, whereas a DC voltage will charge it once and then stop. The overall effect is that a capacitor can pass any changes in voltage signals, as shown in Figure 2.7.

Figure 2.7
An alternating voltage can pass through a capacitance

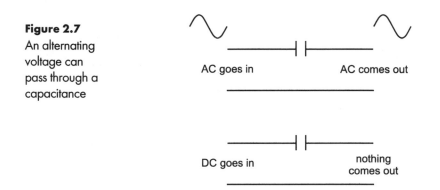

AC goes in   AC comes out

DC goes in   nothing comes out

There are three formulas that may be useful:

The charge stored

$Q = CV$, where Q is the charge in coulombs, C is the capacitance in farads (F) and V is our friend the voltage. If the voltage V changes, then the charge changes so electrons flow into or out of the capacitance.

The energy stored

W or $E = 0.5CV^2$. W or E is energy measured in joules (1 joule per second = 1 watt), C and V are as above.

Reactance

$$X_c = \frac{1}{2\pi f C} \, \Omega$$

The X is reactance (in ohms) and the small 'c' means capacitive. The f is frequency in hertz (cycles per second).

Magnetism is much the same

Electromagnetism

When any current is flowing, the movement of the electrons creates a magnetic field. As current flows along a cable, the magnetic field acts in a circular pattern as in Figure 2.8.

Figure 2.8

A single wire has a circular magnetic field

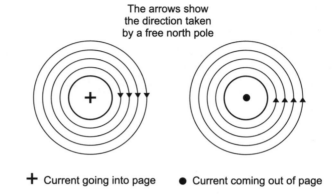

The arrows show the direction taken by a free north pole

+ Current going into page   ● Current coming out of page

When a current is changing, the associated magnetic field is also changing and a changing magnetic field causes a voltage to be created in any nearby conductors. This effect is called inductance and is measured in the unit 'henry' (H). This is the same Mr Henry that we met in Chapter 1. An inductance of 1 H would result in a voltage of 1 V being generated by a current changing at the rate of 1 A per second.

We come across two types of inductance, which can sometimes seem confusing but things are simpler than they seem. If the current changing in a circuit causes a voltage in another circuit, we call this effect mutual inductance (M) but if the voltage is caused in the same circuit, it is called self-inductance (L). The good news is that the formulas are just the same.

Magnetic materials

Even the electrons inside a material are spinning and moving about and so create magnetic fields. In most materials, the magnetic fields are random and have the effect of canceling each other out, and so the material is considered to be non-magnetic. There are some materials where the magnetic effects just happen to align themselves and add together to give a strong external magnetic effect. These magnetic materials are iron, nickel, cobalt, chromium and manganese.

Some magnetic materials are permanently magnetized and are called magnetically hard. This is what we get if we buy a magnet. Some lose their magnetism easily and are called magnetically soft. This is like the core of a coil or transformer that magnetizes easily as electricity flows in the wire and becomes demagnetized as soon as the current stops.

Important effects of magnetism

Any flow of current always creates a magnetic field. Any change in a magnetic field always creates a voltage in a conductor and the polarity of the voltage always opposes the change of current.

There are a couple of formulas that may be useful:

The energy stored

W or $E = 0.5LI^2$. W or E is energy measured in joules.

Reactance

$X_L = 2\pi fL \Omega$, where X_L is inductive reactance in ohms, the f is frequency in hertz (cycles per second) and L is inductance in henrys.

Choice of cable materials

The job of a conductor is to allow current to flow, so the lower the resistance the better. Unfortunately, life is never that simple. As well as low resistance, we would like it to have other characteristics.

Ideally, the best conductor would have zero resistance, zero corrosion, zero weight and be unbreakable. Three characteristics, none of which are achievable, and that's before we start on cost. In real life, we have to see which of these characteristics are most important for the job in hand.

If we are really worried about resistivity then we would go for silver, as it has the lowest resistance, at room temperature, of any material. The others, starting with the best, would be ranked in the order: copper, gold, aluminum and steel.

If corrosion is the biggest problem then we would choose gold, though for large-scale use the cost may be a problem. In order, the others are: silver, copper, steel and aluminum.

Weight may well be a worry, in which case we would go for aluminum followed by steel, copper, silver and finally gold.

And what about strength? The strongest is steel, then comes copper, silver, gold and weakest of all is aluminum. Some apologies are due here to the aluminum suppliers, as the figures used are for aluminum and not aluminum alloys, some of which can exceed steel for strength.

It really is a matter of choosing the material that best suits the job in hand.

To allow the current to be contained inside a cable, the conductors must be enclosed in an insulator. They are nearly all plastic-based these days and we will be looking at them in more detail in Chapter 7.

Effects of temperature

Generally speaking, whether we are looking at insulators or conductors, they perform worse as the temperature increases.

Have a go at these ...

Chapter 2 quiz

1   At room temperature, the best conductor of electricity is:

(a) gold.
(b) silver.
(c) lead.
(d) aluminum.

2   A watt is:

(a) a word used to indicate a question.
(b) the unit of energy.
(c) a unit of inductance.
(d) the unit for the rate of using energy.

3   1000 nanofarads (nF) are the same as:

(a) 1000 pF.
(b) 1 μF.
(c) 1×10^6 MF.
(d) a billionth of a farad.

4   As the temperature decreases, most conductors:

(a) and insulators show a decrease in resistance.
(b) show a decrease in resistance but insulators show an increase in resistance.
(c) show an increase in resistance but insulators show a decrease in resistance.
(d) and insulators show an increase in resistance.

5   **If the current flowing along a cable increases from 2 A to 4 A, the energy stored in the surrounding magnetic field:**

(a) decreases to one-quarter of its original value.
(b) increases to twice its original value.
(c) decreases to half of its original value.
(d) increases to four times its original value.

3

How cables work

Whether we are using cables to transfer power or to pass data, telephone or television signals, we are still transferring power from one place to another. As we send power it must spend much of its journey contained in the cable. This means that the characteristics of the cable must have some direct influence on the data and how well it is transferred so it may be worth our while to spend a few minutes looking at some characteristics of cables.

In Chapter 7 we will return to look at some specific cable designs.

Using cables to transfer power

One of the simplest uses of cable is to connect a power supply to a simple load.

Have a look at the battery connected to a variable load as in Figure 3.1. We are going to see how the load added to the cable can affect the efficiency of power transfer. To do this we change the value of the load resistance from 0 to 10 ohms (Ω) and, at each step, calculate the power available to the load. We will increase the resistance 2 Ω at a time, just to keep the maths easy.

The total circuit resistance is the internal resistance of the power supply added to the resistance of the variable load. We can then find the

Figure 3.1
The resistance increases but will the power increase?

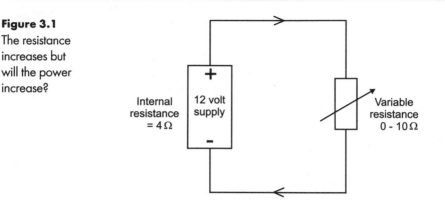

current flowing by using the Ohm's law formula $I = \dfrac{V}{R_{TOTAL}}$ Ω and finally finding the power in the load by using $P = I^2 R_{LOAD}$ watts (W).

Just as an example, when the load resistance is at 6 Ω, the total resistance is $4 + 6 = 10$ Ω. This gives a current of $I = \dfrac{12}{10} = 1.2$ amps (A) and a power in the load of $1.2^2 \times 6 = 8.64$ W. It is easy but tedious to work out the power in the load at any value of load value.

In Figure 3.2, the power transferred to the load is shown for all the different load values.

Figure 3.2
For maximum power the load must be matched

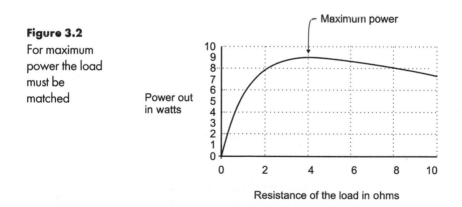

Matching

The maximum power is transferred when the internal resistance of the supply is equal to the resistance of the load. This doesn't only apply in this example, it is a general answer that this is true for any source and

23

any load. When the load has the correct value, it is said to be matched.

When signals are sent over long cables, the input signal has to transfer its power to the cable and, at the far end, the cable has to pass its power on to the receiving equipment. So for efficient power transfer, the source must be matched to the cable and the cable must be matched to the load.

This suggests two questions – what is the resistance of a cable, and what happens if the cable and load are not matched? We will go for cable resistance first.

A quick look at a cable under direct current (DC) conditions

The simplest form of cable is basically two conductors, kept apart by an insulator. It has resistance along the cable because the conductor is not perfect. There is also a slight leakage between the conductors because the insulator is not perfect either.

A short length of cable could be represented by three resistances. R1 and R2 represent the resistance of the copper of the out and return line and R3 is the resistance of the insulator (see Figure 3.3). The total resistance in this example is near enough to 1 MΩ. Theoretically, of course, it is 1 Ω + 1 Ω + 1 MΩ.

If we add another section of cable, as shown in Figure 3.4, the total resistance will have halved to about 500 000 kΩ if we ignore the series resistances. This is because the new bit of cable is in parallel with the 1 MΩ of R3.

Figure 3.3
A short section of transmission line

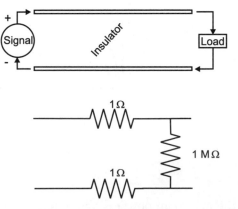

Please note: these values do not represent any particular cable

Figure 3.4
The input
resistance is
now smaller

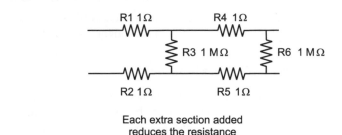

Each extra section added
reduces the resistance

As we add more sections, the small series resistances like R1, R2, R4 and R5 will be trying to increase the resistance and the parallel ones like R3 and R6 will be trying to decrease the resistance. The final resistance will smooth out to a final value, as in Figure 3.5, which we call the characteristic resistance of the cable, the value of which depends on the design of the cable. Curiously enough, if we assume the far end of the cable is short-circuited, a short length will start at nearly 0 Ω but as the length of the cable increases, the series resistances add up to increase the final resistance of the cable. With a very long cable, the short-circuit and open-circuit versions converge on the characteristic resistance.

Figure 3.5
The final
resistance
doesn't change
however long
the cable

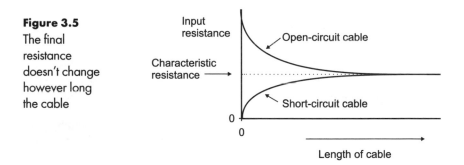

A twin pair, as used in domestic telephone circuits, has a characteristic resistance, called R_0, of 600 Ω.

The interesting thing about the characteristic resistance of the cable is that once the cable is long enough for the resistance value to get close to its final value, then the resistance does not change regardless of how long the cable becomes.

So, what about a short cable?

We cannot take half an inch or a centimeter of open circuit twisted-pair telecom cable and expect it to have an impedance of anything like

600 Ω. From our work above, we can only expect a very high value of resistance, in our example a little over 1 MΩ.

So what do we do with short lengths?

Figure 3.5 suggests that it is difficult to use short lengths of cable until the resistance value homes in on the final characteristic resistance. We can work around this problem by terminating the cable with a load that is equal to the internal resistance. This has the effect of bringing the cable resistance down to the characteristic resistance almost immediately so that even the shortest cable is usable.

Thinking back to the results we had in Figure 3.3, we had a cable resistance of virtually 1 MΩ but if we put a terminating load of 600 Ω across the end as in Figure 3.6, the cable resistance will fall to approximately 600 Ω. Don't reach for your calculator – it's actually 601.64 Ω but it is near enough there.

Figure 3.6
The terminating resistance equals the characteristic resistance

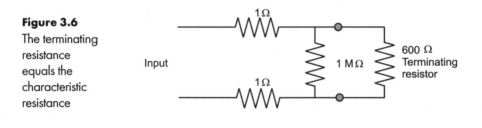

What about signals?

When the cable is used to pass alternating current (AC) signals, particularly at high frequencies, we have to take inductance and capacitance into account.

As the current flows along the cable it gives rise to a changing magnetic field and an inductance. The changing voltages likewise create electric fields that result in a capacitance between the conductors. Just as a matter of interest, these effects are shown in Figure 3.7 but it is hardly worthwhile working through pages of calculation just to prove a figure that the manufacturer will be pleased to give us free of charge.

The opposition to current flow due to inductance or capacitance is called reactance rather than resistance but they are both measured in ohms. The total opposition to current flow is due to a combination of reactance and resistance and is called impedance.

The AC equivalent of characteristic resistance is called characteristic impedance, still measured in ohms and has the symbol Z_o.

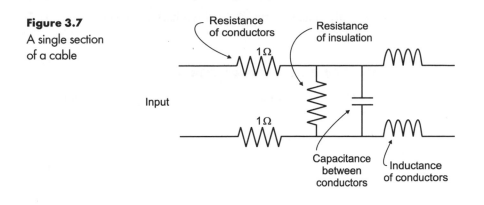

Figure 3.7
A single section
of a cable

At high frequencies, the characteristic impedance is largely dependent on the values of inductance and capacitance and these are quoted by manufacturers as 'values per mile or kilometer'. Whichever lengths are used, we can slip them into the same formula

$$Z_o = \sqrt{\frac{L}{C}} \text{ ohms (approx.)}$$

where L = henrys per unit length and C = farads per unit length.

What happens if the load is not matched to the cable?

Let's start with an extreme case of the line being left disconnected at the far end.

We apply a burst of signal and it starts to travel along the line. Now, moving down the line is both current and voltage, in other words, a small burst of power. The amount of this power is VI watts, or voltage (volts) × current (amps) = power in watts.

At the far end there is an open circuit, so no current can flow. This is significant because we can have no power in the open circuit load because current = 0, and so $VI = 0$ W.

So what happens to the power? It cannot just disappear, it cannot be absorbed by the open circuit load, so the only alternative is that it bounces back along the line. It will return all the way to the transmitter and may cause mischief when it gets there. The 'Great Northeast black-out' that disconnected power from 30 million people in the USA for 13 hours in 1965 started when a trip disconnected on the output from the Niagara Falls generating station. Within two and a half seconds, the resultant reflected energy caused other trips and isolated the 1800 MW generator and from then on it was all downhill.

A very similar thing happens if we short circuit the far end of the cable. If we have a perfect short circuit, all the power is reflected because we

27

cannot have any voltage across 0 Ω. If we had any voltage across 0 Ω, we would have an infinitely high current and this is not possible.

We don't get perfect open circuits or short circuits

No, that's right. Apart from fault conditions occurring at the end of the cable, in most cases we are dealing with slight mismatches in values.

To calculate the proportion of power being reflected back along the cable we need to know the characteristic impedance Z_o and the terminating impedance Z_t and once we are there, it's number-crunching time.

Here is the formula:

Reflection coefficient $= \rho = \dfrac{Z_t - Z_o}{Z_t + Z_o}$ (ρ is the Greek letter rho)

A short-circuit termination would make $Z_t = 0$ so we would get:

$$\rho = \frac{0 - Z_o}{0 + Z_o} = \frac{-Z_o}{+Z_o} = -1$$

and so the reflection coefficient is -1. The '1' tells us that all the power is reflected and the minus sign tells us that the voltage at the output is reduced, in this case, totally canceled.

An open circuit termination would make $Z_t = \alpha$ (infinitely high), so the top line would be $\alpha - Z_o$, which is equal to α, and the bottom line would also be equal to α, so the value of $\dfrac{\alpha}{\alpha} = +1$. This means total power reflection, and the '+' means that the voltage at the termination is increased.

So, let's see what happens if a mismatch is neither completely open circuit nor short circuit.

Worked example

A 100 Ω line is terminated by 50 Ω causing a mismatch. What is the reflection coefficient?

Feed the numbers into the formula and crank the handle:

$$\rho = \frac{Z_t - Z_o}{Z_t + Z_o} = \frac{50 - 100}{50 + 100} = \frac{-50}{150} = -0.33$$

Another example

A 100 Ω line is terminated by 99 Ω causing a mismatch. What is the reflection coefficient?

$$\rho = \frac{Z_t - Z_o}{Z_t + Z_o} = \frac{99 - 100}{99 + 100} = \frac{-1}{199} = -0.005$$

Return loss

In the above examples, we can see that the reflection coefficient is getting smaller as the mismatch is getting smaller.

The return loss (R) is the ratio of the reflected voltage to the transmitted voltage. This ratio is always less than one and is called the reflection coefficient.

$$\text{Return loss, } R = 20 \, \log \left(\frac{\text{reflected voltage}}{\text{transmitted voltage}} \right)$$

Because the reflected voltage is always less than the voltage we send down the line, the value of the log is always negative.

The better the matching, the lower the reflected power and the higher the negative value of the return loss. This can be a little confusing as a 'good' value of return loss is numerically larger than a 'poor' value. This is hard on the brain so we will work through a couple of examples to see how the numbers crunch.

Baddy – the first example included a bad mismatch as we terminated a 100 Ω line with 50 Ω and had a reflection coefficient of 0.33.

The return loss is $R = 20 \log(0.33) = 20(-0.48) = -9.6 \, \text{dB}$

Goody – the second example is a slight mismatch with a reflection coefficient of −0.005.

The return loss is $R = 20 \log(0.005) = 20(-2.301) = -46.02 \, \text{dB}$

Numerically, 46.02 is larger (and better) than 9.6 (forget about the − sign).

A popular misconception about current flow

Although current is the movement of electrons, it is quite wrong to think that electrons in a cable actually move along the length of the cable at 60–90% of the speed of light, which is 3×10^8 m per second (186 000 miles per second). In fact, the electrons in a conductor travel quite slowly in the direction of the current flow, they drift along at about one meter per hour (0.0006 miles per hour).

But when I turn the light on, I don't have to wait 10 minutes for the electricity to get to the bulb. So how does current flow so fast? It is the effect of the movement of the electrons that is rapid. Imagine a tube full of balls, as in Figure 3.8. If we were to push another ball into one end, then almost instantaneously a ball would fall out of the far end. It is very easy to assume that the ball has traveled at lightning speed through the tube. It is only the effect that has moved from one end to the other extremely rapidly.

29

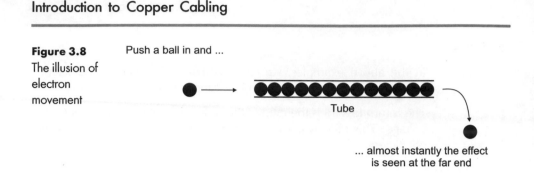

Figure 3.8
The illusion of
electron
movement

Push a ball in and ...

Tube

... almost instantly the effect
is seen at the far end

Alternating current is even stranger. The electrons in a 60 hertz (Hz) supply only move for 1/120th of a second before they reverse in direction and so travel only tiny distances. Hence, the actual electrons in your electric heater are the same electrons that were there when it was built. The utility companies do not supply our houses with electrons but they just provide the power to move our own electrons to and fro in the wire. An amazing thought.

Some magnetic effects

Skin effect

If we pass an AC current along a conductor, a magnetic field is created that expands from the center of the conductor, as shown in Figure 3.9.

Figure 3.9
The magnetic
effect of an
increasing
current

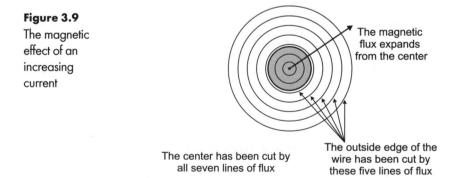

The magnetic
flux expands
from the center

The center has been cut by
all seven lines of flux

The outside edge of the
wire has been cut by
these five lines of flux

We mentioned in the previous chapter, the way that a changing magnetic field induces a voltage in any conductor – even in the same conductor. Now, the higher the frequency the faster the current changes, the faster the magnetic field strength changes the more voltage that is induced. If we just look at the conductor for a moment, we can see that as the current changes its center has been cut by more flux than its outer edges and so there is more voltage induced in the center.

Remembering that the induced voltage always opposes the current, there will be more opposition in the center than on the surface. What happens now is that the current flowing in the conductor finds it easier to flow along the surface than in the center. Have a look at Figure 3.10. To some extent this occurs with all conductors but we take it into account when high frequencies are used combined with high powers as in commercial transmitters.

Figure 3.10
The current tends to flow on the surface due to skin effect

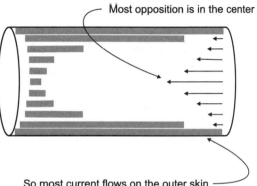

Most opposition is in the center

So most current flows on the outer skin

To carry high powers we need thick copper conductors and once we have seen that the center of the copper carries very little current it was just a small step to decide to reduce the amount of middle by having a flat tape which increases the amount of 'outside' and decreases the 'middle'. We can also decide not to have a middle at all and use a lighter, self-supporting copper tube.

Skin effect is noticeable around 25 MHz but becomes very significant over 1 GHz. At this frequency we can silver-plate the conductors and the skin effect is so pronounced that it restricts the flow of current to the very surface region and nearly all the current flows through the silver layer. So we get all the benefits of the current flowing through the best conductor without the cost of solid silver conductors.

Electromagnetic interference (EMI)

Every electronic device that we use involves current flow and signals. Now it is an inescapable law of nature that any changing current creates a changing magnetic field and, even worse, any changing magnetic field creates a voltage in all surrounding conductors. This means that it is inevitable that our television and our computer are transmitting interference signals over the surrounding areas.

An example of interference between cables is 'crosstalk' where the signal from one cable can be detected in others nearby. There will be a bit more on this in Chapters 7 and 11.

EMI? Relax – you probably won't notice it

The vast majority of EMI has no effect on us or our equipment. Much of it is too weak to cause a problem. Every time we switch on our television, we create interference by radiating a signal and so does every other television in the world.

So how come we can still use our television?

Two reasons. EMI gets weaker as it moves away from the source, so to receive signals from a television 20 miles away is quite possible but we would need to build a very sensitive receiver. Quite possible, after all we receive tiny transmissions from spacecraft like *Voyager* when millions of miles away and with much less power than a television.

The more sensitive our equipment, the more problem we have with EMI and the less sensitive, the less of a problem it is. Something very insensitive like a vehicle starter motor never fails or accidentally runs due to interference.

The second reason is that televisions do not interfere with televisions because they are not designed to respond to the frequency of signals that other televisions produce. We are not bothered by high-frequency pulses transmitted by bats because they are way outside of our hearing range. We cannot see infrared light, so this radiation may not bother us but may irritate fish that can sense it.

We will be back to have another look at EMI and its prevention in Chapter 7.

Chapter 3 quiz

1 **Which of these is the best reflection coefficient?**

   (a) 1.
   (b) +100.
   (c) 0.
   (d) −100.

2   If a cable has a characteristic impedance of 100 Ω, doubling the length of the cable would result in an input impedance of:

(a) 100 Ω.
(b) 50 Ω.
(c) 200 Ω.
(d) −100 Ω.

3   Skin effect:

(a) cannot occur in insulated cable.
(b) is more noticeable at very high frequencies.
(c) is caused by static electricity.
(d) is more noticeable at very low frequencies.

4   Maximum power can be transferred when the load impedance:

(a) and the characteristic impedance of the cable and the source impedance are all equal.
(b) is 75 Ω.
(c) is as high as possible.
(d) is equal to $\sqrt{\dfrac{L}{C}}$ Ω.

5   The speed of light is:

(a) 186 000 m per second.
(b) 1 m per hour (0.0006 miles per hour).
(c) 3×10^8 m per second.
(d) 300 million miles per hour.

4

Decibels – they get everywhere but what are they?

Every time we look in a catalog or read about anything technical, we come across decibels. The decibel is one of those things that other people know about and sound most impressive when they talk about them. But nobody actually tells us what they are.

The two ways of doing decibels

We can skip straight to the answer and not worry about all the details. This is tempting providing that one of our readymade answers fits our problem. The other way is to devote an hour and be able to understand them forever so we can solve all problems for ourselves.

In this chapter we will tackle both methods. We'll look at the quick and easy method first.

Getting by with decibels

Decibels are a way of stating the amount of gain or loss of power or voltage that occurs in the circuit.

We pick up a telephone and take for granted that our voice can be heard next door or in any part of the world even without shouting! Our voice signal gets smaller as it travels along the cable and before it disappears altogether we have to pass it through an amplifier to make it larger again. The reinvigorated signal sets off on its journey once again.

The amplifier must strengthen the signal by the same amount that the cable weakens it. If we look up the data on the amplifier, its gain would be expressed in decibels as would the losses on the cable.

Bigger or smaller?

Just look at the number of decibels in the data book. If it has a '+' in front of the number of decibels, or it says 'gain', 'amplification' or something like it, then the signal is getting larger.

If the number of decibels starts with a '−' sign, or mentions 'loss' or 'attenuation', then the signal is getting smaller.

What if we have a truck load of amplifiers and lengths of cable?

In a situation like Figure 4.1 we can just add them up, so if we had an amplifier with a gain of 20 decibels (20 dB) and we used a cable with a loss of 15 dB to connect it to another amplifier with a gain of 10 dB and another length of cable with a loss of 12 dB then we could look at the whole circuit. Using + and − for gains and losses, the circuit would contain + 20 − 15 + 10 − 12 = +3 dB, so overall it would have a gain of 3 dB. Note the abbreviation for decibels: small 'd', capital 'B', and never put an 's' on the end.

Figure 4.1
What is the overall gain or loss in decibels?

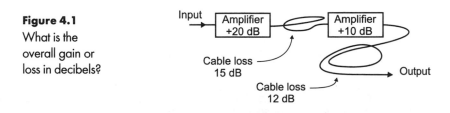

There is no limit to how many gains and losses that we can add.

What do the numbers mean?

They are based on the ratio between the output power to the input power. We have to be a bit careful here because the number of decibels quoted depends on how the measurements were taken. If we measured the power at the input and output we would get a different result than if we took voltage or current measurements.

The reason for this is that the formula used to find power is

$$\frac{\text{voltage}^2}{\text{resistance}}$$

or current^2 × resistance if we measure currents. Both of these terms use squared terms, either voltage or current, and these have the effect of doubling the numerical value of the decibels.

Just as an example, if an amplifier doubles the input power, we could quote this as having a power gain of 3 dB. If we took the same amplifier but compared the voltage at the output with the voltage at the input, the result would be 6 dB, so in adverts or data books, we need to be alert to how the measurements were taken.

Some useful numbers

We have just seen that a power gain of 3 dB doubles the power. And 6 dB doubles it again, and 9 dB doubles it again and so on. The same thing works for voltages or currents except they double after 6 dB, and again at 12 dB and at 18 dB, and so on forever.

Rather than calculate all the figures, we can go straight to Table 4.1 for the answers. Here are some examples of its use.

Table 4.1 Decibels

dB gain or loss	Power ratio gain	Power ratio loss	Voltage gain	Voltage loss
1	1.26	0.79	1.12	0.89
2	1.58	0.63	1.26	0.79
3	2.00	0.50	1.41	0.71
4	2.51	0.40	1.58	0.63
5	3.16	0.32	1.78	0.56
6	3.98	0.25	2.00	0.50
7	5.01	0.20	2.24	0.45
8	6.31	0.16	2.51	0.40
9	7.94	0.13	2.82	0.35
10	10.00	0.10	3.16	0.32
20	100.00	0.01	10.00	0.10
30	1000.00	0.001	31.62	0.03
40	10000.00	0.0001	100.00	0.01
50	100000.00	0.00001	316.23	0.003
60	1000000.00	0.000001	1000.00	0.001

Example: what is a power loss of −7 dB?

A power loss of 7 dB would reduce the power at the output to 0.2 of the input power.

Example: what is a voltage gain of +8 dB?

A voltage gain of 8 dB would result in an output voltage that is 2.51 times greater than the input voltage.

Example: how many decibels do these voltages represent?

The input signal level is 18 volts (V) and the output is 30 V. Use Table 4.1 to estimate the decibel equivalent.

Output voltage/input voltage for the amplifier is 30/18 = 1.67.

In the table, 4 dB = 1.58 and 5 dB = 1.78, so 1.67 is somewhere in-between 4 dB and 5 dB. The disadvantage of tables like this is that they cannot give very accurate results for the actual circuit we are working on.

The alternative is to invest an hour in learning to calculate the decibel values rather than using a table. The maths looks awkward at first glance but there are only two formulas that we ever use; one for power and the other for voltages or currents.

The decibel is a logarithmic unit

The word 'logarithm' is usually abbreviated to 'log', which sounds much friendlier although still a bit of a worry. Let's play around with a few numbers just to get some terms sorted.

If we start with the number 100, we can also write this as 10^2 (or 10 squared).

But how would we describe the number 2 in the number 10^2?

It is called the logarithm or log of 100, so the log is just a fancy name for the power of 10.

How to find the log of any number

Ten cubed or 10^3 = 1000, so the log of 1000 is 3.

But what about a number like 350 that is between 100 and 1000? It seems sensible that the log must be somewhere between two and three. It is, but we need some help to find the number.

Reach for a calculator. On the calculator:

Press the log button.

Punch in the 350.

Hit the "=" button, and the answer is given as 2.54, if we ignore some of the decimal places.

We can say that $10^{2.54} = 350$ or put more simply, the log of 350 is 2.54.

The log values don't often work out to nice clean numbers so it's quite usual to round off the final log value. While we have our calculators ready, there are a few log values that give interesting results that it may be better to see now rather than later. Try them first and we can have a look at the results in a moment.

1. Have a look at log 10. This is the way we normally write 'log of 10'.
2. What about log 0.4?
3. What is the log of 1?
4. Try finding the log of zero.
5. Finally, try the log of −350.

1. log 10 = 1. This is because $10 = 10^1$.
2. log 0.4 = −0.39 (roughly). Notice how all numbers less than 1 give negative log values.
3. Zero. This is because $10^0 = 1$.
4. and 5. There are no log values for zero or any negative number. They don't exist. The calculator will probably give an error warning.

Having found a log, how do we work back to find the number?

We know that the log of 200 is 2.301, so $200 = 10^{2.301}$.

To find the value of $10^{2.301}$ just use the 10^x button on a calculator. Different calculators have different methods for doing this but very often the 10^x button is marked on the same button as the log function that we have been using. In this case, we press the 'shift' button then press the log button. To find the value of $10^{2.301}$ the complete button sequence is ... shift, log, 2.301, =

Note the slight rounding off error that has crept in. This is normal, just forget it.

We can use logs to multiply and divide numbers

Let's take an easy example. Say we want to multiply 100 by 1000 to get the obvious answer of 100 000.

The log of 100 is 2, the log of 1000 is 3 and the log of the answer is 5, so we have added the numbers. By using the calculator to find the value of 10^5, we get our answer of 100 000.

Now we can do the same trick to divide 100 000 by 1000 to give the answer 100. The log of 100 000 is 5, the log of 1000 is 3 and the log of the answer is 2 so we have subtracted the logs of the numbers. Finding the value of 10^2 gives us the answer of 100 as expected.

Summary of logs

- To multiply numbers, add their logs.
- To divide numbers, subtract their logs.
- Logs cannot be used to add or subtract numbers.

Back to decibels

The decibel is used to compare the power at the output of something to the power at the input. It is a logarithmic unit so it obeys all the rules of logs.

The formula for a power gain or loss in decibels is:

$$\text{power gain or loss} = 10 \ \log\left(\frac{\text{power}_\text{out}}{\text{power}_\text{in}}\right) \ \text{dB}$$

Using decibels with an amplifier

An amplifier has a higher output power than its input power, so it is said to have a power gain, as shown in Figure 4.2.

Figure 4.2
What is the gain
in decibels?

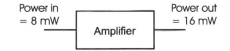

Power in
= 8 mW

Amplifier

Power out
= 16 mW

Once we know the power level going into the amplifier and the level coming out, all we have to do is to put the figures into the decibel formula as shown below. In this case, we can see that the power at the output is greater than that at the input, so we have a power gain

$$\text{gain} = 10 \ \log\left(\frac{\text{power}_\text{out}}{\text{power}_\text{in}}\right) \ \text{dB}$$

Insert the power values into the formula:

$$\text{gain} = 10 \ \log\left(\frac{16 \times 10^{-3}}{8 \times 10^{-3}}\right) \ \text{dB}$$

Calculate the bit inside the brackets, which in this case comes to 2:

$$\text{gain} = 10 \ \log 2 \ \text{dB}$$

Use a calculator to find the value of log 2, which comes to about 0.3. We are now left with

gain = 10 × 0.3 dB

Multiplying it out gives a final gain of 3 dB.

Worth remembering: a doubling of the power level = 3 dB.

Summary

To find the number of decibels:

Start with the formula: $\text{gain} = 10 \log \left(\dfrac{\text{power}_\text{out}}{\text{power}_\text{in}} \right)$ dB.

- Find the value of the bit inside the brackets.
- Take the log.
- Multiply by 10.

Decibels and attenuators

If the output power is less than the input, then a loss has occurred. We also say that the signal has been attenuated. If the circuit has been designed to produce a known degree of attenuation, we call it an attenuator. A loss has occurred in Figure 4.3.

Figure 4.3
Decibels with losses

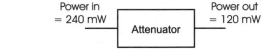

Power in = 240 mW — Attenuator — Power out = 120 mW

Whether we are dealing with an amplifier or attenuator, we always use the same formula to calculate the number of decibels:

$$\text{gain} = 10 \log \left(\frac{\text{power}_\text{out}}{\text{power}_\text{in}} \right) \text{dB}$$

Insert the power values into the formula:

$$\text{gain} = 10 \log \left(\frac{120 \times 10^{-3}}{240 \times 10^{-3}} \right) \text{dB}$$

Calculate the bit inside the brackets, which in this case comes to 0.5:

gain = 10 log 0.5 dB

Use a calculator to find the value of log 0.5, which comes to about −0.3 (notice the minus).

We are now left with

gain = 10 × −0.3 dB

Multiplying it out gives a final gain of −3 dB.

Worth remembering: just as we have seen that 3 dB is a doubling of power, a halving of the power level = −3 dB.

If the decibel value is positive, a gain has occurred. If the decibel value is negative, a loss has occurred.

Mind your language

The results with this attenuator can be described in two ways and it is very important that we don't get them confused.

The result of the calculation was −3 dB so it would be OK to say the circuit has a loss of 3 dB. We could also say that it has a gain of −3 dB. In the first case, the word 'loss' will emphasize that a loss has occurred and in the second case, the minus sign will also make it clear.

What we must avoid is saying that the loss was −3 dB, as this sentence is open to two different interpretations due to the double negative. The best description would be 'a loss of 3 dB'.

How we use decibels in a real circuit

In Figure 4.4, what is the total gain or loss in decibels?

Figure 4.4
What is the total gain or loss in decibels?

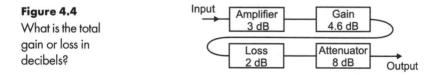

Input → Amplifier 3 dB — Gain 4.6 dB

Loss 2 dB — Attenuator 8 dB → Output

Looking at the circuit, we can see two increases of power, one referred to as 'amplifier' providing +3 dB and the other as 'gain' giving +4.6 dB. The names used don't matter – they mean the same thing. The 'loss' and the 'attenuator' also cause a reduction in power. The change in power levels is one of −2 dB and the other of −8 dB.

To find the total change round the circuit we just have to add up all the values: +3 + 4.6 − 2 − 8 = −2.4 dB. The circuit causes an overall loss of 2.4 dB.

If the input power to this circuit is 20 watts (W), what would be the output power?

41

In Figure 4.5, the circuit has been grouped together to show the single loss.

Figure 4.5
What is the
output power?

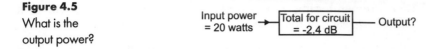

Input power = 20 watts → Total for circuit = -2.4 dB — Output?

Some more maths – but it's not too bad

We have previously used the standard formula to find the number of decibels when we knew the input and output powers. This time, we want to work the other way – we have the decibels and we want to find the power.

1. Write down the standard formula

$$\text{gain} = 10 \log \left(\frac{\text{power}_{out}}{\text{power}_{in}} \right) \text{dB}$$

2. We start by entering the gain in decibels. We have the gain (−2.4), so it now looks like this:

$$-2.4 = 10 \log \left(\frac{\text{power}_{out}}{\text{power}_{in}} \right) \text{dB}$$

3. Divide both sides by 10

$$-0.24 = \log \left(\frac{\text{power}_{out}}{\text{power}_{in}} \right) \text{dB}$$

4. To get rid of the log, we can find the antilog of both sides. On the right-hand side, it just cancels the 'log' term but on the left we have to use the 10^x button on the calculator as we did a few pages ago. The value of $10^{-0.24}$ is 0.5754 or roughly 0.58:

$$0.58 = \left(\frac{\text{power}_{out}}{\text{power}_{in}} \right)$$

5. The input power was 20 W so we can pop this in the formula:

$$0.58 = \left(\frac{\text{power}_{out}}{20} \right)$$

6. Just a bit of tidying up now by multiplying both sides by 20. This will give the result:

$$0.58 \times 20 = \text{power}_{out}$$

So the output power is 11.6 W.

Summary

To find the output power:

Always start with the basic formula: $\text{gain} = 10 \log\left(\frac{\text{power}_{out}}{\text{power}_{in}}\right)$ dB.

1. Enter the gain (or loss).
2. Divide by 10.
3. Use 10^x to find the antilog.
4. Put in the value of the input power.
5. Multiply both sides by the input power.

All done.

What if wo know the gain and the output power but don't know the input power?

Figure 4.6 shows the problem. This is a very similar problem to the last one so, unsurprisingly, the solution follows much the same pattern.

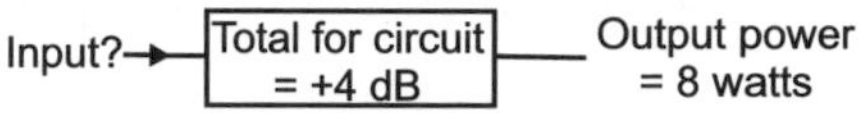

Figure 4.6
Find the output power

Input? → | Total for circuit = +4 dB | → Output power = 8 watts

1. Always start with the same formula:

$$\text{gain} = 10 \log\left(\frac{\text{power}_{out}}{\text{power}_{in}}\right) \text{ dB}$$

2. Enter the gain in decibels:

$$4 \text{ dB} = 10 \log\left(\frac{\text{power}_{out}}{\text{power}_{in}}\right) \text{ dB}$$

3. Divide both sides by 10:

$$0.4 = \log\left(\frac{\text{power}_{out}}{\text{power}_{in}}\right) \text{ dB}$$

4. Use the 10^x button on the calculator to convert the decibels to a gain ratio. The value of $10^{0.4}$ is 2.5 (this also removes the 'log' term):

$$2.5 = \left(\frac{\text{power}_{out}}{\text{power}_{in}}\right)$$

5. Write in the 8 W output power:

$$2.5 = \frac{8}{\text{power}_{in}}$$

6. We have to fiddle this around to get the input power. This is done in two steps. Firstly by multiplying both sides by power$_{in}$ to give 2.5 × power$_{in}$ = 8 and then by dividing both sides by 2.5 and we get:

$$power_{in} = \frac{8}{2.5} = 3.2 \text{ W}$$

So we have an input of 3.2 W. As a final check, we can always use the output and input powers to calculate the gain in decibels. These figures come out to 3.98 dB but there are always rounding off errors with decibels, so it's probably OK.

Summary

To find the output power:

Start with the formula: $gain = 10 \log \left(\frac{power_{out}}{power_{in}}\right) \text{ dB}$.

1. Enter the gain (or loss).
2. Divide by 10.
3. Use 10^x to find the antilog.
4. Put in the value of the input power.
5. Multiply both sides by the input power.

All done.

Using decibels as a power level

My signal generator states its output power is decibels. This seems odd. We have spent the last hour going on about decibels being a ratio between the input and the output of a circuit, but the signal generator has no input signal.

The trick here is to pretend that we have an input power level. This works if we all agree on what input power level is going to be assumed. The assumed level is normally 1 mW and to let everyone know that we are using this level, we add a small 'm' to change the decibel symbol from 'dB' to 'dBm'. For completeness, we should also say that in telephony this 1 mW is measured at a frequency of 1 kHz across 600 ohm (Ω).

Once we got the idea of putting a letter after the 'dB' symbol, it started being used for other power levels like 'dBmW' (power relative to 1 mW), 'dBW' (power relative to 1 W) and 'dBmV' (voltage relative to 1 mV across 75 Ω). There are lots of them but we hardly ever meet them.

My signal generator claims an output power of 30 dBm. From Table 4.1, we can see that 30 dB is a power increase of 1000 times. Now, using the 1 mW level as the input, the output is 1 mW × 1000 = 1 W.

I think that's enough on decibels, just have some fun with the little quiz that follows. If there are any problems, all the answers are explained at the back of the book.

Chapter 4 quiz

---

1   An output of −20 dB means that the power has been:

(a) reduced by 10 times.
(b) increased by 1000 times.
(c) reduced by 100 times.
(d) increased to 0.1 W.

---

2   A power level of 30 mW can be written as:

(a) +30 dB.
(b) 14.77 dBm.
(c) +14.77 dB.
(d) 3 dBW.

---

3   If a communication link resulted in a power loss of 6 dB, two such links joined up would cause a loss of:

(a) three-quarters of the input power.
(b) 3 dB.
(c) 12 dB.
(d) 6 dB.

---

4   A system with an input power of 35 mW and an output power of 127 mW has a gain of approximately:

(a) 5.6 dB.
(b) 3.63 dB.
(c) −2.44 dB.
(d) 15.44 dB.

---

5   An amplifier has a gain of 23 dB and an output power of 5 W. The input power is approximately:

(a) 200 mW.
(b) −23 dBm.
(c) 1000 W.
(d) 25 mW.

5

How is data transmitted?

The simplest and most popular system

Just talk to someone! Like all transmission systems, this involves a transmitter, a transmission medium and a receiver. To use this system in its simplest form we have few requirements that, in most cases, are given to us free of charge: a voice to make the sound, air for it to travel through and ears to receive it.

The drawbacks are obvious. We are limited in range. We are limited in the number of messages that can be passed at the same and we cannot send pictures. And it is terribly slow. When we try to communicate with a person who speaks a foreign language, we can immediately add another drawback – the sender and the receiver must be using the same format for the transmissions.

The telephone

We can extend the range of transmission by using a microphone to convert the voice into an electrical signal which can then be sent along a pair of wires to the far end, at which point a loudspeaker of some form converts the signal back into sound.

Two wires are required because the current from the power source must flow out of one terminal, through the circuit and back to the other terminal. It is just about possible to get away with using only a single conductor and using the earth return as the other wire. It works but it usually makes the circuit too noisy to use.

Analog systems

The frequency range of our voice and hearing is usually quoted as 20 Hz–20 kilohertz (kHz), though for adults this is rather optimistic. Like most other things, our hearing gets worse as time goes by. Even so, we don't need the full range to talk. The telephones restrict the range to 300 Hz–3.4 kHz, which is good enough for conversation even though it sometimes makes recognition of a person difficult and, as we all know from experience, totally destroys the quality of music.

The gap between 300 Hz and 3.4 kHz is 3.1 kHz, and this is called the bandwidth of the system. The bandwidth of a system determines how fast information can be sent over a communication link.

Distortion on transmission lines

When signals are being passed along the cable they consist of a mixture of frequencies.

Unfortunately, not all frequencies are treated equally by the cable. As the frequency increases there is a tendency for lower frequencies to be passed more easily that the higher ones, so if we had a signal containing a 2 kHz and a 10 kHz sine wave with the same amplitude, at the far end we would find that the 2 kHz signal would have a greater amplitude than the higher frequency signal, which has suffered more losses. This would distort the sound and is called frequency/attenuation distortion.

At telephone frequencies, this effect is not very noticeable and causes no serious problems.

In a similar way, a cable passes higher frequencies at a faster rate so they suffer phase distortion, as shown in Figure 5.1. In this diagram, the higher frequency component has moved faster than the other one. On the voltage/time graph the faster component appears to have been shifted towards the left of the diagram – in other words it is getting to there sooner. The displacement of this frequency has distorted the resulting waveform. This is not a great problem at voice frequencies but it can be a nuisance at higher frequencies.

If we pop back to Figure 3.7 we will remember that a cable contains capacitance between the conductors and inductance in series with each. Phase and attenuation distortion are caused by the capacitance and to some extent they can be counteracted by increasing the inductance or 'loading' the cable. We do this by connecting a coil (inductor) into the cable at intervals of either 1.3 km (4500 ft) or 1.8 km (6000 ft).

Figure 5.1
Phase distortion

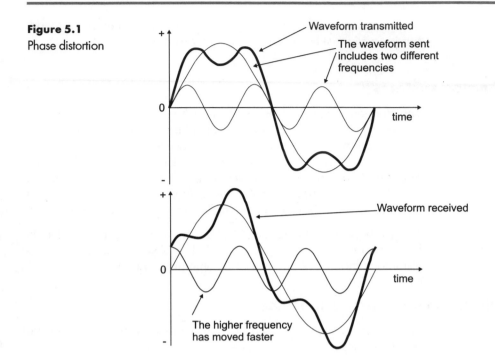

Waveform transmitted

The waveform sent includes two different frequencies

time

Waveform received

The higher frequency has moved faster

time

Multiplexing – or getting more for our money

In the simple analog telephone system, we spoke into a microphone and our voice frequencies would be transmitted along to the speaker at the far end. If someone else tried to use the same connection, we run the risk of our voices becoming hopelessly jumbled together.

One of the ways to send more information down a single connection is to multiplex the transmission. In analog systems we multiplex by using a system called frequency division multiplexing or FDM.

Frequency division multiplexing

By using an electronic circuit called a modulator, we can 'attach' our audio telephone conversation to a higher frequency. If we attach our telephone range of 300 Hz to 3.4 kHz to, say, a frequency of 1 MHz, the resulting frequencies would be between 1 MHz + 300 Hz and 1 MHz + 3.4 kHz, which is 1.0003 MHz to 1.0034 MHz.

Along comes another subscriber wishing to use the telephone cable. No problem, we could attach his conversation to a 2 MHz signal. His frequencies would extend from 2 MHz + 300 Hz to 2 MHz + 3.4 kHz. The resulting frequencies are so far apart that they could easily be separated at the far end with no interference – so we have used the same bit of cable to carry two conversations at the same time.

We could attach more analog data to 3 MHz, 4 MHz and so on, as in Figure 5.2 and we could pack the channels much more closely than the 1 MHz spacing chosen for this example. The only limit is that the channels must be separated sufficiently to prevent interference and the upper limit must be within the transmission capability of the cable design.

Figure 5.2
Sharing a
telephone
cable

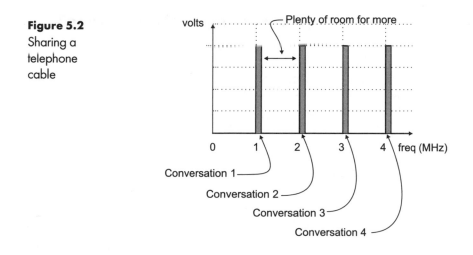

Noise is a serious drawback to analog communication

In an analog system, the amplitude and frequency of the original speech or other signal is transmitted along the cable. If there should be a crackle of lightning or we introduce a hum by passing the cable too close to a power cable then the system simply adds these new noises to the speech. At the far end, we have no way of knowing which parts of the signal are wanted – is it an electrical hum, or is the other person entertaining us by humming a tune? We don't know.

But there is a way to spot these unwanted noises and remove them but, as with everything else in life, it comes with a price tag.

Digital transmission

This is definitely a significant improvement over analog and as new systems are installed, the world will go digital but it will take many years. The old analog methods are installed and working so there is no great pressure to spend all the money on digging them up.

In the analog system the frequency of the signal carries the information about the frequency of the original input, so if our friend has a squeaky voice the signal on the cable will have a higher frequency than someone with a lower-pitched voice and if our friend speaks louder, the amplitude of the signal increases. This means that both the frequency and the amplitude of the received signal are important and this is why we could do nothing about the hum and lightning crackle.

How do digital systems get around the problem of noise?

The digital transmission converts the signal to a series of voltage pulses. These pulses are all of the same amplitude – let's say 5 volts (V). When any interference occurs, the voltage will wobble above and below this level. The good news is that the circuitry can easily check to see if there is a pulse and if there is, we know it should be 5 V so we just change it back to 5 V and the interference has been removed for us. This technique, called regeneration, effectively rebuilds the signal at intervals along the transmission route so that a digital signal sounds as clear if it is coming from the other side of the world as it would if it is coming from a few miles away. Of course it would – the regeneration trick means that the sound we are actually listening to was rebuilt just a few miles away.

The conversion of the original sound into these pulses must be fairly clever so that the frequency and loudness of the sound is included in some way so we know whether our friend is being loud and squeaky or muttering quietly.

The advantages of using digital transmission

1. Noise and distortion. We have already mentioned the way we can clean up the signals to overcome these problems.
2. Security. We can use encryption methods like we do on the Internet to prevent or reduce the likelihood of illegal access.
3. Mixed transmissions. As we can convert any signal to a digital format, a single route can carry voice, video, telegraphy or anything else.
4. Store and Forward. We can easily store digital information in a memory system for transmission at a later time, just as we can with our e-mails.
5. Automatic correction. If we transmit our digital signals in a known pattern, it is possible for the receiving station to be able to recognize an error if one has occurred and, in some cases, to make a correction.

And the price of all these goodies?

Not very much these days. The system cost can be greater but the difference is decreasing as the cost of the electronics continues to fall and the physical installation cost is much the same. The bandwidth of a digital transmission is usually greater and we have to put up with that.

Bits and bauds

Digital signals normally send two voltage levels, maybe 5 V and 0 V, often called high and low or 1 and 0. Having just two levels, they are called binary digits volt and abbreviated to 'bit' from BInary digiT.

Transmission rates

The rate at which binary data are transmitted by a digital communication system is measured in bits per second (b/s).

We sometimes come across the term 'baud'. When we ask what baud means we often get the answer – 'Don't worry about that, it's just the same as bits per second' and often when we look up the specification we do indeed find the two figures are exactly the same.

However, there are occasions where they may be different. Technically, a baud is the reciprocal of the shortest element in a transmission.

What?

If we used a system in which binary information is sent by causing a phase shift in the transmitted signal, we could arrange that a 0° phase shift = 00 in binary and 90° = 01, then 180° = 10 and finally 270° = 11. Now a transmission 'element' would be a single transmitted event and in this case would be a single change of phase and let's assume this occurs in 0.1 second. We could do this 10 times per second and the specification would state the transmission rate as 10 baud. Each phase shift represents two bits, so we are transferring information at the rate of 20 bits per second. In this case, the baud rate and the bit rate are not the same. It doesn't happen very often but it is worth being aware that it could happen.

Preparing for digital transmission

The original analog waveform undergoes a series of conversions before it is ready for digital transmission. It may be a good idea to start with an analog waveform and follow it through the process, so a millisecond's worth of a telephone conversation is shown in Figure 5.3.

51

Figure 5.3
A small part of
an analog
signal

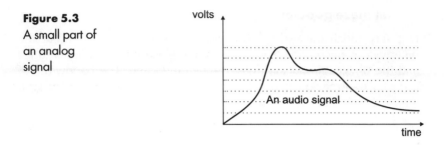

Sampling

We start by measuring the amplitude of the waveform at intervals. If we record its amplitude often enough, we can get enough information to allow us to reconstruct the waveform without transmitting the whole thing.

How often is 'often enough'?

This is the result of a lot of mathematical analysis and is given the impressive name of the Nyquist Criterion. Luckily, it is much easier than it sounds. The end result is that the sampling rate must be at least twice the highest frequency component contained in the sound.

In our telephone system, the highest frequency transmitted is 3.4 kHz so the sampling rate must be at least 2×3.4 kHz = 6.8 kHz. It can be higher than this, of course.

The samples for our example waveform are shown in Figure 5.4. The original analog signal has now been converted to a series of pulses – remember that the dotted outline is just a reminder of the original signal but no longer exists and won't be shown anymore.

Figure 5.4
We have taken
some samples

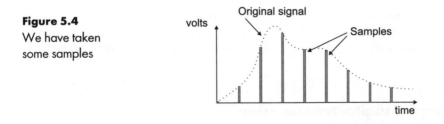

The pulses look like a digital signal but they are not. Sure, they are pulses but the test for a digital signal is that the pulses are restricted to set amplitudes. These samples can take on any value depending upon the amplitude of the analog signal at the time of being sampled. At this time, the signal is called pulse amplitude modulation (PAM), which is still an analog system.

Quantization

This is the time that we finally lose any analog information. The vertical scale measures voltage so we can number it in volts, millivolts or any other interval that seems sensible at the time.

The sample amplitudes are measured against this scale and allotted the nearest number, so if we are working in volts, as in Figure 5.5, the sample may be given the value of 5 V or 6 V but no intermediate values. Obviously, the more values we use, the more accurate the results, so working in millivolts would be a lot more accurate than working in volts. The drawback of using a very large number of steps is that it increases the amount of data that must be transmitted and this means using a greater bandwidth or sending the data more slowly.

Figure 5.5

The voltage of each sample is measured

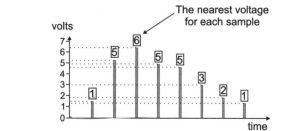

Whatever scale we use, the sample height is still taken to the nearest size on the scale, so an error will often occur unless, by chance, the sample heights are all exactly equal to one of the scale values. These errors are called quantization errors. In Figure 5.5 we can see these errors. We have three samples with a value of 5 but by looking carefully, we can see that the samples are actually different heights.

Pulse code modulation (PCM)

The sample sizes are expressed in binary numbers so that we can transmit the data in a digital format using combinations of 0 and 1. The number of bits used will be a power of two: 2, 4, 8, 16, 32, etc. and to reduce the size of the quantization errors a popular choice is 256, which is represented by an 8-bit number.

To transmit each sample level we will need to transmit an 8-bit number, so if we want to send 1000 samples in a second and each requires 8 bits to be sent, our digital transmission will be running at a rate of 1000 × 8 = 8 kb/s = 8000 bits per second.

The transmitted signal now contains no analog information at all. At the far end we can reverse the process by converting the groups of bits back into voltages and rebuilding the samples. Finally we can reconstruct the original waveform and send it to the receiver.

Time division multiplexing (TDM)

You may remember that we used frequency division multiplexing in an analog system to allow us to send many telephone signals down a single cable. In digital systems we can play a somewhat similar trick.

As there are time gaps between the samples, we can pass samples of other waveforms along the cable during these unused times. A simple mechanical example is shown in Figure 5.6 but in a real circuit we use high-speed electronic switching using a clock signal to keep them in step.

Figure 5.6
Four signals
along one
cable by TDM

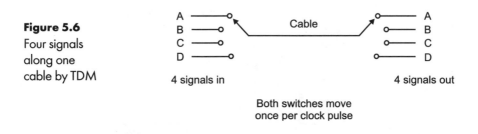

With the switch in the position shown, the data from circuit A is connected through and a single PCM sample is transmitted to receiving circuit A. A moment later, both switches move one position and circuit B is connected to send its sample and similarly with circuits C and D. The switches then return to position A and it all starts over, as in Figure 5.7.

Figure 5.7
Samples on a
TDM system

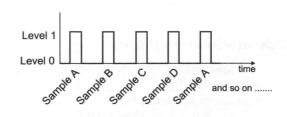

One sample from each, then back to the start

The importance of synchronization

Assuming that the switches move at the same instant is easier said than done. If one switch was not activated on one occasion, then instead of connecting A to A, B to B, C to C and so on we may get out of step. If the second switch missed a go, we may get the sequence A to A, B to A (stuck!), C to B (going again), D to C, then back to A to D, B to A, C to B, D to C and so on. This would be a calamity.

What would happen? Your customer A could find all its data going to its rival headquarters at company D. This would make D happy until it realized that all its data was going to company C, B's data would end up in A and C's data would go to B and our communication company would lose several clients.

It is important that we keep the electronic clocks at each end of the system running in synchronism for two reasons: to ensure that the data levels are read at the correct moment so that we receive accurate data and to avoid the wrong data being sent to the customers, as mentioned above.

We could do this by using an extra cable to carry clock pulses but this would be a waste of resources – we could be using this extra cable to carry more data for paying customers. Instead we use the sudden changes in voltage levels as we send the PCM signals to snap the receiving clock back into synchronism so that each bit of data is accurately read by detecting the data at the right moment.

This cures the first problem but what about the second? What if the communication link is disconnected for a second or two – how do we identify the destination for each block of data?

Framing

One way of solving this is by framing. In this case, each burst of data is preceded by a block of data which can be recognized by the receiving circuits to indicate the correct destination. This is just like writing an address on an envelope so that the right letter gets to the correct address.

A voltage problem

If all the signals varied between, say, 0 V and +5 V then we may have an average voltage of about 2.5 V on the line. By saying 2.5 V we have assumed that the 0 V and +5 V signals are turning up evenly mixed. But what happens if the signal doesn't change for a while, just like before. The average direct current (DC) voltage will wander up and down, which can cause electronic problems at the receiving end as the signals are being converted back to an analog signal.

55

Figure 5.8 shows the problem of loss of synchronism when we don't have enough voltage level changes and the troubles caused by the DC voltages but we can cure both of these by some electronic trickery called encoding.

Figure 5.8
Two problems
that need
solving

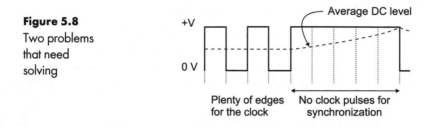

Encoding

There are many systems of encoding that are in use, each with some useful features for particular jobs, but by looking at three common versions we can see the possibilities. If we ever need details of the other types, we can consult the reference books.

Basically, we have the two problems: we want to keep plenty of edges to synchronize the clocks and we want to stop the average voltage from wandering about.

Figure 5.9 shows the codes: the NRZ, Manchester and AMI methods of encoding. The employment of these forms will be discussed in Chapter 9.

NRZ (non return to zero)

This is the basic system in which we have just the two levels, 0 and 1. This has the disadvantage of having very few changes of level and so we could have long periods of constant level with no edges for synchronization and the DC levels can easily wander.

Manchester

Each sample slot is split into two halves. The first half carries the real data that is read by the receiving circuit and the second half carries an inverted version. So when we want to send a 0 sample, we actually send 01, and to send a 1, we send 10. This method means that there will always be an equal number of zeros and ones and so the average DC voltage level does not wander about. It also cures the clock problem by generating truck-loads of edges even when the input signal stays at a constant value.

Figure 5.9
Some examples
of encoding

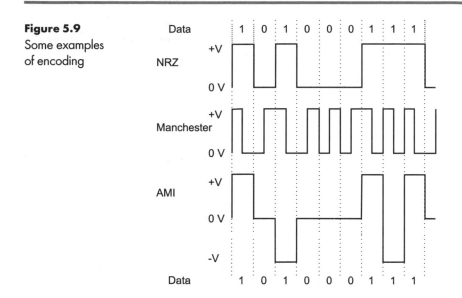

AMI (alternate mark inversion)

In this one, zeros remain at zero levels but alternate level ones are inverted. This balances out the DC voltage level and generates synchronizing edges when ones are being sent but unfortunately, a series of zeros do not produce any edges.

Chapter 5 quiz

1 When sampling a transmission which includes frequencies between 2 and 6 kHz, the lowest acceptable sampling frequency is:

(a) 2 kHz.
(b) 4 kHz.
(c) 6 kHz.
(d) 12 kHz.

2 The encoding method that generates the greatest number of edges is:

(a) NRZ.
(b) Manchester.
(c) AMI.
(d) double inversion.

3   As the frequency of transmission on a cable increases the:

(a) losses decrease.
(b) speed of transmission decreases, causing distortion of the signal.
(c) signals are unaffected.
(d) speed of transmission also increases.

4   Quantization error can be reduced by:

(a) using a larger number of levels.
(b) using higher frequency input signals.
(c) using a more accurate input signal.
(d) increasing the spacing between levels.

5   When converting analog into a digital format the order of events is:

(a) TDM, sampling, quantization, PCM.
(b) TDM, PCM, sampling, quantization.
(c) sampling, quantization, PCM, TDM.
(d) sampling, quantization, TDM, PCM.

6

We don't do it like that

The problem of standards had to be tackled but few people would think of regulations, standards or codes as being interesting. There are, of course, some people that greet each new regulation with yelps of delight but they are not ordinary folk.

Having overcome the temptation to ignore standards and such things altogether or to put them all at the end of the book hidden away as an appendix, it seemed better to write the rest of the book and leave this chapter until last and see where it would be most useful. In the end, six seemed to be a good number.

What are codes and standards?

They are recommendations and sometimes legal requirements and sometimes recommended ways of designing communication systems or methods of working.

The difference between a requirement and a recommendation is not very great since, as a professional, we are expected to conform to 'good practice' and failure to do so could make us responsible for any loss or damage.

Codes are regulations and rules backed up by national laws that are intended to provide safety during installation and use of materials. Standards, on the other hand, are devised to ensure that the system installed actually works and continues to work for a reasonable period of time. It means that we can buy equipment from different suppliers and be sure that they will be compatible when they are installed.

Are they a good thing?

Yes. They are good for all of us. Customers, installers, designers, repairers – in fact it's a no-lose situation.

Imagine a situation in which a person has wired all the communications in a building using a non-standard method – perhaps the installer had a warehouse full of non-standard blue-cored cable. If all cores were blue and a fault developed whoever came along to repair it would have a serious problem – even the original installer would probably find it quicker to re-work the whole system.

Where do they come from?

They grew by personal recommendation into local customs and from there to national standards. The national standards have to some extent been outgrown by the increasing globalization of the world.

There are hundreds, probably thousands, of committees, usually staffed by volunteers, working on standards throughout the world. Some of these actually formulate the regulations or recommendations and some act as bodies to encourage cooperation to provide more widely recognized standards and codes. The workforce is very mobile and both materials and designs can be purchased and exchanged on an international basis, particularly now that the Internet makes all countries effectively the same distance apart.

Alongside the national standards, the move is towards uniformity over larger areas such as Europe. National and area standards are converging towards the day when we have a single world standard.

A small corner of the organization behind the production of a standard in shown in Figure 6.1.

National organizations and standards

Despite the trend towards globalization, the national codes and standards still take precedence over the international agreements where they clash. This allows for local customs and language differences. Quite often there are differences in the technical terms adopted in various parts of the world.

The organizations fall into three groups. There are the bodies that actually produce the codes and these are supported by others that help to

Figure 6.1
A sample of the organizations behind the standards

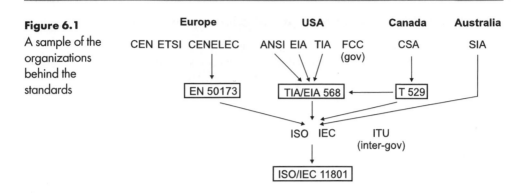

coordinate the work and connect with international and government bodies. Lastly, there are inputs from the various governments.

USA

ANSI (American National Standards Institute – www.ansi.org)

Like many of these organizations, this is a private not-for-profit concern that provides a forum and assists other organizations to reach agreement but does not impose its own solutions. It was founded in 1918 and its membership includes about 1500 companies, government bodies and international representatives. It is also the only US representative on the ISO, the world body.

EIA (Electronic Industries Alliance – www.eia.org)

Starting life as the Radio Manufacturers Association, it now represents 80% of the American electronic industry with a membership of 2300 companies employing 2 million people.

It has worked with TIA and ANSI to produce the standard ANSI/TIA/EIA 568, which is now the most popular telecommunications standard in use in the USA in our field.

TIA (Telecommunications Industry Association – www.tiaonline.org)

This is a trade organization of about 900 companies that provides assistance to their members in selling products and services and such things as market analysis.

It also cooperates with the TIA and ANSI in producing industrial standards. Their members manufacture and distribute almost the complete range of electronic products available in the world.

FCC (Federal Communications Commission – www.fcc.gov)

This is an independent US government agency set up to regulate inter-state and international communications. It reports directly to the US Congress. It is directed by five commissioners appointed directly by the US President.

Europe

CEN (Europe Committee for Standardization – www.cenorn.be/)

Based in Belgium, this is a European organization to promote voluntary technical harmonization within Europe in conjunction with other world-wide bodies. Every country within Europe has its own national standards body, in the UK it is the BSI (British Standards Institute). It covers a wide range of standards but for our purposes it particularly works in partner-ship with the two organizations that follow.

ETSI (European Telecommunication Standards Institute – www.etsi.org)

This institute stands alongside CENELEC to produce the European stan-dards for telecommunications. It is based in France and has 54 member countries.

CENELEC (Europe Committee for Electrotechnical Standardization – www.cenelec.org)

This organization representing 19 European countries works within CEN to produce European standards within the area of electrotechnology and, in conjunction with ETSI, telecommunications. It cooperates with ANSI, TIA and EIA in working towards further harmonization of standards.

The European standard is called EN50173 and is very similar to ANSI/TIA/EIA 568. There are a few differences due to preferred terminology and local customs but most of these differences are minor and will be discussed as and when they arise.

Australia

Standards International Australia Ltd (www.standards.com.au)

This limited company represents Australia on the international standard committees.

Canada

CSA (Canadian Standards Association – www.csa.ca)

This organization acts as a third party to consider any suggested change to present standards or the development of a new standard. It is the usual not-for-profit type of organization.

Wherever it appears to make sense, it is harmonized with North American and international standards. Its present standard CSA T529 is harmonized with the American ANSI/TIA/EIA 568. It contributes directly to the international bodies, the ISO and the IEC.

Global

There are three bodies involved, two basically commercial and one governmental.

ITU (International Telecommunication Union – www.itu.int)

Based in Geneva, Switzerland, the ITU is an international organization within which governments and commercial organizations can coordinate world-wide telecommunication networks and services. The ITU, rather like ANSI, does not devise its own standards but helps its 189 member countries to reach a consensus for the benefit of all.

IEC (International Electrotechnical Commission – www.iec.ch)

This body prepares and publishes international standards for all electrical, electronic and related technologies. The committees here take information from the national and regional bodies together with the intergovernmental organizations. It cooperates with the ISO to produce the final international standards.

ISO (International Organization of Standardization – www.iso.ch)

This is a collection of 140 national standards bodies based in Geneva, Switzerland. It is slowly organizing the whole world, producing thousands of standards from paper sizes like A3, A4, etc. to the sizes of credit cards and freight containers. It has also produced the management standards ISO 9000 and 14000 that lay down rules for 'quality management' and 'environmental management'.

In the field of cabling, it has inputs from the IEC and has produced an international standard called ISO/IEC 11801, which is similar, but not identical, to the American ANSI/TIA/EIA 568 and the European CENELEC EN 50173.

Real standardization is getting closer.

In the next chapter, we will have a closer look at some real cables.

No quiz time here – we deserve a rest.

7

Not all cables are the same

Any cable consists of a conductor or several conductors surrounded by an insulator. The conductor is usually copper or occasionally aluminum. They may have outer protective layers to prevent damage. And that's about it. End of book!

As always, it isn't quite that easy. Let's have a look at a few types.

Power cables

These vary according to the amount of current that is to be carried. The more current it has to carry, the lower the resistance of the conductor needs to be. This is because a significant amount of the power being distributed can be wasted in heating the cables. Power = current^2 × resistance, or as a formula $P = I^2R$ watts, where P (sometimes called W) is in watts (W), I is current in amps (A) and R is resistance in ohms (Ω).

Since power is also equal to volts × amps, we can send the same amount of power by making the voltage larger and the current smaller. This means less power loss in the transmission system. It is for this reason that the utility companies use hundreds of thousands of volts for power distribution.

Even with these high voltages, we still have significant currents and to keep the resistance as low as possible, the thickness of the cable is fairly large – like 100 sq. mm or mm^2 (4 sq. inches).

These cables are obviously very much larger than the ones we use for telecoms. When looking at cable sizes we often come across the American Wire Gauge. This seems a little confusing at first because the largest numbers are actually the thinnest wires and none of the measurements are actually nice round numbers. All this is just a result of the way that the copper wires were made and is not just designed to irritate us.

American Wire Gauge

When copper wire is made, it is drawn through a series of dies which makes the wire thinner (and longer of course). In the American Wire Gauge (AWG), after passing through the first die it has a diameter of 7.34 mm (0.289 inches). This size wire is then called 1 AWG.

After passing through the next die, it is reduced in diameter and the size is quoted as 2 AWG and so on. The most popular sizes that we use in telecoms are listed in Table 7.1.

Telecommunication cables

Glancing through the telecom catalogs there seem to be pages of different cables but when we look more closely we see that there are only a few basic designs but many slight variations.

Once installed, cables last a long time. This is both a good thing and a problem. This means that we will meet many cables that range from this year's latest creation to last year's flavor and many examples of older technology that are limping along, designed to meet standards that are no longer applicable. On the other hand, it is nice to know that they don't wear out once a year, otherwise the cost of replacements would

Table 7.1 Wire diameters

AWG	Diameter (mm)	Diameter (inches)	Weight (kg per km)
19	0.91	0.036	5.8
22	0.64	0.025	2.9
23	0.57	0.023	2.3
24	0.51	0.02	1.8
26	0.4	0.016	1.14

A couple of notes: The weight is just for the bare copper. The diameter is given for solid wire – stranded wire comes up slightly thicker because of the slight gaps.

be astronomic. It does mean that we have to be aware of previous cable standards to allow us to recognize the system that we are upgrading.

Whether new or old – there is one problem that all the cables have.

Electromagnetic interference (EMI) and the common cable cures

You may remember from Chapter 3 that when varying magnetic fields cut any conductor it always creates or induces a voltage into it. This is the way the universe works and we can't stop it. In some cases it is a good thing since it allows transformers and televisions to work but in other cases it is a nuisance. We do have some defense methods available.

Blocking it out

This is done by preventing the electromagnetic waves reaching the conductors.

To protect cable from EMI, we just have to wrap it up in a conductor like aluminum or copper – this we can do in several different ways. We can use copper or aluminum foil, or metallized plastic tape or quite often we use copper or aluminum braid for flexibility and lightness. Very occasionally we use a solid sleeve.

The idea behind it is based on our two basic effects of magnetic fields – a moving or changing field always creates a voltage and hence a current in a conductor – and this current always creates a magnetic field that is opposite to the original magnetic field. The incoming EMI creates a current in the covering which creates a field equal and opposite to the EMI and thus the two fields cancel and the cable cores are not cut by the EMI and are protected.

We have concentrated on electromagnetic interference as this has proved to be the main problem that we face. However, we do sometimes suffer from electric fields like the ones that we can feel when we bring a charged balloon near to our hair, and rightly scare us in a thunderstorm. The good news is that the same foil or braid can protect the cable against electric fields just by connecting the ends of the cable to a good ground point. Any electric charge can then leak away.

Types of cable

Stranded or solid

In each type of cable, we generally have the choice of stranded or solid construction.

Stranded is more flexible and will stand being moved repeatedly and is therefore ideal for the final connection to equipment and adjustable links or patch cords between telecomm equipment or circuits. However, stranded cables suffer from increased losses when compared with solid cores and so are less suitable for long lengths.

Solid core will work-harden and break if continuously flexed but it is better at conducting electricity and is therefore best used for the fixed parts of the installation.

Connectors are usually designed for either stranded or solid core, so it is best to read the instructions before use.

There are only two basic types of cables used and the first is coaxial, like we use on our television aerials and with our video recorders, and the other is just pairs of insulated wires.

Coaxial cables

Coaxial cable has a central insulated conductor which may be a solid wire or stranded. It is then enclosed in a conducting layer which is usually a copper or aluminum mesh or sometimes with a solid metal sleeve. It is then covered by an outer insulator called a jacket. The earthed braid provides a barrier against EMI moving into and out of the coaxial cable.

The central core and the outer sheath share the same axis, hence they are coaxial and the cable is referred to coaxial cable or more usually just 'coax' (see Figure 7.1).

Figure 7.1
Coaxial cable

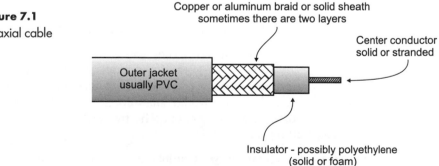

Copper or aluminum braid or solid sheath
sometimes there are two layers

Center conductor
solid or stranded

Outer jacket
usually PVC

Insulator - possibly polyethylene
(solid or foam)

It can carry more data than the twisted pair that we will look at in a moment but is generally more expensive – there are many different propriety standards – and is not easy for general use. Compared with other cables, it is rather bulky and is more difficult to install as it cannot bend so easily.

Coax is widely used for video and television but is not recommended for carrying data. The new standards are likely to continue its decline and possible demise.

Twinaxial, triaxial and triple coax

We sometimes come across the terms 'twinaxial' and 'triaxial'. Despite the impressive names they are really quite simple. They are both basically coaxial cables, the only differences are that twinaxial has two conductors running through the center of the coax and triaxial has a single central conductor but a double layer of braid or other screening conductor. To complete the confusion, some catalogs also list 'triple coax cable', which has a single screen but three separate conductors.

Grounding

Coaxial cable screens should be earthed at just one point. Leaving it disconnected (floating) or grounding it at each end can make the EMI worse rather than better. Further details are given in Chapter 12.

Transfer impedance

This is used to measure how well a shield (screen) is doing its job. We apply some EMI and measure the resulting currents induced into the screen. Then we see how much of the signal can penetrate the shield by measuring the voltage in the center core. The ratio of the voltage to the current is the transfer impedance – the lower the result the better. Zero would be nice!

Twisted cable – over 100 years old and still doing fine

It is true with all EMI that the closer you are to the source of the interference, the worse it is. Now if we could arrange that the two conductors in a transmission pair are always equally distant from the source of the EMI then they would suffer equal levels of EMI and the effects would balance and cancel out. This twisting method is effective up to about 30 megahertz (MHz).

The signals are carried on two conductors – one carrying the signal to the receiving circuits and the other providing the return path. The total 'out' current and the 'return' current are the same, so these cables are referred to as 'balanced', unlike the coax cable.

We refer to cables not in terms of the number of conductors but in terms of the number of pairs starting with a single pair and extending up to large numbers like 4200 pairs. Each conductor is color-coded for easy recognition, otherwise it would be a total nightmare. More about the color codes soon.

Unshielded twisted pair or UTP

This is done by twisting the two conductors. We call this type a twisted pair. We have made no attempt to apply any shielding to block the EMI and so it is often called unshielded twisted pair or UTP.

In Figure 7.2 we can see that, by twisting the wire, each conductor spends the same amount of time close to the source of interference and therefore has the same value of induced voltage. In Figure 7.3, the voltages are seen to be acting in the opposite direction around the circuit and the induced voltages will be trying to push the current in opposite directions around the circuit and their effects will cancel.

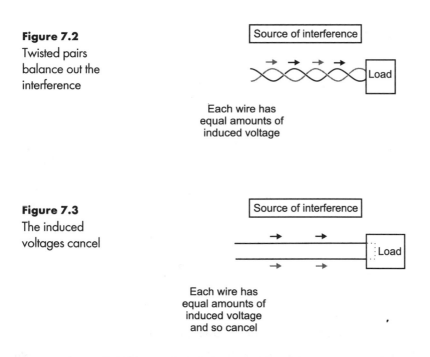

Figure 7.2
Twisted pairs
balance out the
interference

Source of interference

Load

Each wire has
equal amounts of
induced voltage

Figure 7.3
The induced
voltages cancel

Source of interference

Load

Each wire has
equal amounts of
induced voltage
and so cancel

It is very important to keep the wires twisted right up to the point of connection, otherwise the voltage will not completely cancel and shield the connectors at the ends of the cable.

The twisted cable is not always the victim in the EMI wars – it is sometimes the aggressor as it radiates its own EMI to cause problems with all surrounding cables and equipment. Luckily, twisting the wire works just as well to stop its own interference as it did when we were trying to stop the incoming EMI.

As we would expect, the explanation is just like playing a film in reverse. We start off with the interfering circuit as in Figure 7.4 and then in Figure 7.5 we see the effect of twisting the wires. Once the wires are twisted, the conductors are continuously changing their positions so the radiated signals from one length of cable are acting in opposite directions to the signals from the next length. This has the effect of producing equal and opposite electromagnetic fields and significantly reduces the amount of EMI that can be radiated, or received.

Figure 7.4
The current in the cable is radiating interference

Figure 7.5
Twisting the wires greatly reduces EMI

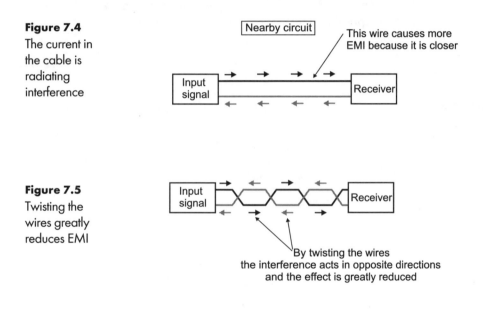

At high frequencies, we can no longer rely on the twisting alone to save us from EMI and we have to add a further layer of defense.

Screened twisted pair (ScTP), also called foiled twisted pair (FTP)

When we are using a twisted-pair cable in situations where EMI is likely to be a problem or the cable is likely to radiate more than we wish, we can add a screen around the cable immediately under the outer jacket. The screen is usually copper or aluminum foil wrapped around the bunch of conductors.

ScTP includes four twisted pairs within an overall shield. It has a characteristic impedance of 100 ohms (Ω) and conductors of diameter 0.51 mm (24 AWG). An ScTP cable is shown in Figure 7.6.

Figure 7.6
Screened
twisted pair
(ScTP) or foiled
twisted pair
(FTP)

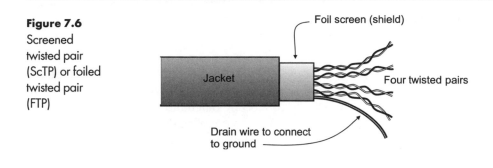

Shielded foiled twisted pair (S-FTP)

This cable consists of four twisted pairs surrounded by a common metallic foil shield followed by an outer braided metallic shield.

Shielded twisted pair (STP) or fully or double-shielded twisted pair (SSTP)

This has the same shield under the jacket as we saw in ScTP but, in addition, it has a separate screen around each pair, as shown in Figure 7.7. This provides improved protection against external EMI and also protects against crosstalk between each pair and allows the cable to operate at high frequencies up to 600 MHz. We will have another look at crosstalk in Chapter 11.

STP cable is not interchangeable with the ScTP we saw in the last section.

Figure 7.7
Shielded
twisted pair
(STP)

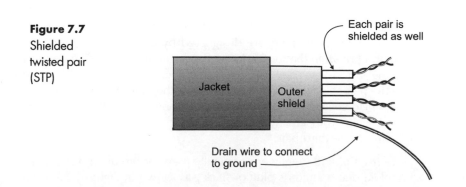

A few general bits

Balun

When passing signals from a twisted-pair cable to a coaxial cable, we are connecting a balanced system to the unbalanced transmission of the coax cable and usually connecting different characteristic impedances. To make this transition, we use a coupling device called a 'balun'. This name is a contraction of 'BALanced UNbalanced'.

Drain wire

To provide an easy way of grounding it, an uninsulated copper wire is included to make good contact with the foil throughout the length of the cable and it also gives us an easy way to make the earth connection at each end of the cable.

I've found a nylon drain wire!

Since nylon is an insulator, this is not likely. This is a plastic 'Slit-Wire' or 'Rip Cord' which, when pulled, can cut along the outer jacket to make its removal easier.

A useless bit of knowledge – nylon was developed jointly in New York in the USA and London, UK (NY + LON = nylon).

Recognition

To prevent any confusion, the jacket of cables must be marked to identify the type of cable. For example, a ScTP cable would be marked with '100 Ω ScTP' in addition to any other information required by local or national codes.

Color-coding

Four twisted-pair cables have their conductors colored in four pairs for easy recognition. The pairs are designated by the colors blue, orange, green and brown in that order. In each pair the first one, often referred to as the 'tip', is basically white with a stripe, band or something else using the pair color. The second member of the pair, called the 'ring', is just a single solid color. So if we saw a conductor with an insulator colored white with green stripes, we can recognize it as the first or 'tip' wire in the third pair.

The terms 'tip' and 'ring' were originally used for the tip and ring parts of a telephone operator's plug or 'jack', as shown in Figure 7.8.

Figure 7.8
The origin of the
'tip' and 'ring'

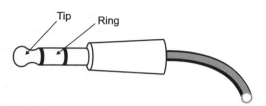

Table 7.2 Pair colors

Pair number	Tip	Ring
1	White/blue	Blue
2	White/orange	Orange
3	White/green	Green
4	White/brown	Brown
5	White/slate	Slate
6	Red/blue	Blue
7	Red/orange	Orange
8	Red/green	Green
9	Red/brown	Brown
10	Red/slate	Slate
11	Black/blue	Blue
12	Black/orange	Orange
13	Black/green	Green
14	Black/brown	Brown
15	Black/slate	Slate
16	Yellow/blue	Blue
17	Yellow/orange	Orange
18	Yellow/green	Green
19	Yellow/brown	Brown
20	Yellow/slate	Slate
21	Violet/blue	Blue
22	Violet/orange	Orange
23	Violet/green	Green
24	Violet/brown	Brown
25	Violet/slate	Slate

Cables with more pairs need more colors and these are designated up to 25 pairs and are shown in Table 7.2. The colors for the four pairs are the first ones listed.

Need more than 25 pairs?

If we have more pairs, each set of 25 pairs are wrapped up separately by 'binders', which are colored threads or tapes used to identify them.

Categories and classes

In the last chapter we saw how the American ANSI/TIA/EIA team and the international ISO/IEC group were creeping closer together. The standards are converging but there is a slight difference in terminology when looking at the cable performance as the US standards use the term 'Category' whereas the International group uses 'Class'.

The American categories range from 1 to 7 at the moment, with 1 being the slow end of the scale whereas the International classes run from A to F, with A being the slow end. There are slight technical disparities between the regulations, so there is a little more to it than just changing letters and numbers. Some of the differences will be mentioned as we look at the categories.

For new work the recommended categories are 3, 5e, 6 and 7 but since we will be meeting other grades in dealing with existing installations, it is worthwhile to spare a few moments to look at all the grades.

Category 1 and Class A using UTP cables

A very low-speed cable sending data at less than 100 kHz. It can handle analog voice transmissions and ultra-slow data like alarm signals and switch control. It is very cheap but is seldom installed as it is so limited and is not recognized by the current standards.

Category 2 and Class B using UTP cables

Also obsolete but may still be in use for low-speed data such as digital voice transmission. Again, not recognized by the current standards. Cat 2 has an upper limit of 4 MHz but Class B only extends to 1 MHz.

Category 3 and Class C using UTP cables

These cables are recognized by the current regulations and are current usage with cables specified up to 16 MHz. The cable referred to are four-pair twisted cable or larger core numbers bundled in groups of 25 cores.

During installation, Cat 3 cable should not be untwisted more than 75 mm (3 inches), otherwise we get too much EMI.

This category and class are typically used for a wide range of applications within the limits of the 16 MHz bandwidth. It takes care of a lot of things and is installed in a large proportion of the current installations but, with a view to future use, new installations may prefer to use a higher grade and Cat 5e is recommended.

Typical uses are digital and analog voice, Ethernets and any LANs (local area networks), ISDN (integrated services digital network) and DSL (digital subscriber line).

Category 4 using UTP cable

This has been scrapped. It is no longer recognized by current standards. It was only slightly better than the Cat 3 cable and cost much the same as Cat 5 and 5e. It never really caught on because it was superseded before it had established itself in the market.

Categories 5, 5e and Class D using UTP, FTP and S-FTP cables

These cables are designed to support applications requiring bandwidths up to 100 MHz.

Cat 5 was the most popular cable for new installations and 5e is really just a Mark 2 version with additional transmission characteristics that support applications that use all four pairs for simultaneous bi-directional communication. With an eye on future use, it is now recommended that Cat 5e is now adopted for all new installations and Cat 5 will fade away.

Class D always includes a specification for fiber-optic cable, which is not included in the current ANSI/TIA/EIA-568-B specs.

Because of the higher frequencies being used we have to be more careful about untwisting the cable and causing too much EMI. For this reason, Cat 5e cable should not be untwisted more than 13 mm (0.5 inch).

Category 6 and Class E using UTP or ScTP cables

Cat 6 is specified to 250 MHz and this is just about the highest frequency for UTP cable and ScTP is likely to the best choice despite its drawbacks of cost and some additional difficulties in installation – grounding and increased bend radii. The screen should be grounded at each end, providing that ground voltage levels are not greater than 1 volt (V), otherwise we can get excessive current in the screen conductor.

Category 7 and Class F using STP (SSTP) cables

Cat 7 is designed to operate at frequencies up to 600 MHz and is beyond the reach of unshielded twisted pairs. A filler often runs through the center of the cable to keep the pairs apart. This is typically a soft plastic strip which is X-shaped to separate the cable space into four compartments each holding a cable pair.

Who mentioned Cat 8?

No-one, at least not many people. It is likely that things will head towards fiber optics when Cat 7 starts to look a bit slow. But it was as recent as 1999 that Cat 7 was seen as unlikely ever to be considered. But with telecom speeds doubling every two years and computer speeds doubling every three years – what speeds are we looking at in 12 years' time? Our computer and our cables may well be running at over 30 GHz. Nonsense, we say – but someone probably said that when we were installing Cat 1 cable and someone thought about sending signals through glass 'wires'.

It's time to see what catalogs say about cables and other things . . .

Chapter 7 quiz

1   The highest cable category that can use unshielded cable is:

(a) 4.
(b) 5.
(c) 6.
(d) 7.

2   Select the correct statement from those below:

(a) a copper wire of 24 AWG is twice as thick as a 12 AWG wire.
(b) using the AWG, a 14 gauge wire is thinner than a 12 gauge.
(c) a solid core copper core has a higher resistance than a stranded core.
(d) a stranded copper core is only used with a solid shield.

3   In a four-pair cable described as ScTP:

(a) each pair of conductors is separately screened but there is no outer screen.
(b) each pair of conductors is screened and there is an overall screen enclosing all pairs.
(c) a twisted foil runs through the center of the cable.
(d) there is an outer screen under the jacket but the pairs are not screened.

4   FTP cable is most like:

(a) ScTP.
(b) SSTP.
(c) STP.
(d) UTP.

5   New installations should use categories:

(a) 3, 5, 5e or 7.
(b) 1, 3, 5e or 6.
(c) 4 or 5e.
(d) 3, 5e, 6 or 7.

8

Selecting, protecting and connecting cables

Selecting a cable

The first step is to study the technical requirements of the job in hand and see how they match up with the categories or classes of cables that are available. We must also bear in mind compatibility with any existing cabling and possible future developments. No customer ever comes back to ask for the system to handle less traffic at a lower speed.

Different makes can be different

All Category 5e cables meet Cat 5e standards, of course, but the standards only state the lowest possible levels of performance that are acceptable and not the highest. So whether we are buying a burger, a car or a cable we are always going to face a range of prices and qualities even though they all meet the appropriate requirements.

Reputable cable companies will always be happy to provide the specifications for their cables so rather than skip over the specs page it would be worth comparing them.

Cable specifications

Here are the main specs that are likely to be supplied. To comply with their stated category, they must at least equal the required figures over a range of frequencies. Sometimes a supplier just says 'it complies' and others list their measured results at intervals across the frequency range. Where figures are quoted in this chapter they refer to examples found in typical Cat 5e cables.

Frequency range

It will certainly quote a range that is at least equal to the category requirements but will sometimes exceed it. The bigger the better.

Structural return loss (SRL)

Back in Chapter 3 we looked at return loss where we sent a signal along a cable and when it reached the end, part of the energy may be reflected due to a mismatch in the impedance of the load compared with the characteristic impedance of the cable.

Structural return loss is just like this but it is a measure of the cable performance and not the load matching. We send a signal down the cable and measure only the reflections that occur from slight changes within the length of the cable. We express the ratio of the transmitted power to the reflection in decibels (dB).

A typical cable may have values of about 20 dB, which means about 1% of the signal. The value of the return loss is often least at high and low frequencies, and peaks somewhere in the mid-frequency range.

Attenuation

It is inevitable that power losses will occur as the signals pass along the cable as the current flows through the resistance of the conductors. There are also losses due to electromagnetic interference (EMI).

Wait a moment – if a cable is screened, we prevent EMI so surely we can't have losses due to EMI.

Unfortunately, this is not so. As the current flows it creates a magnetic field. The magnetic field cuts the copper screen and induces a voltage in it. This voltage causes currents to flow in the screen and these currents result in a power loss (I^2R watts). This power must have come from the power we are feeding into the cable and is therefore a loss and contributes to the power loss and the attenuation.

To measure the attenuation of a cable, we first have to ensure that the end of the cable is matched so we don't have any reflections to spoil the readings. We first measure the voltage at the near end, before the power enters the cable and then take a similar voltage reading at the far end just before the load.

Crunch the numbers using the formula:

$$\text{Attenuation} = 20 \ \log \left(\frac{V_{in}}{V_{out}} \right) \text{dB}$$

The attenuation figure is normally expressed as dB/100 m.

If we had a perfect line with no losses, the value of V_{in} and V_{out} would be the same and the attenuation would come out to 0 dB, so we are hoping for a result that is as low as possible.

The attenuation is least at the lowest frequency and a typical cable Cat 5e cable may be 2 dB/100 m at 1 megahertz (MHz) and rise to 22 dB/100 m at 100 MHz.

Near-end crosstalk

We met crosstalk (Xtalk or XT) in Chapter 3 as unwanted transfer of signals from one pair of conductors to another and in cable specifications we come across four variations on the theme.

We will start with near-end crosstalk (NEXT), which is shown in Figure 8.1. In this case, we have applied a signal to one end of a pair and then

Figure 8.1

Measuring near–end crosstalk (NEXT) and far–end crosstalk (FEXT)

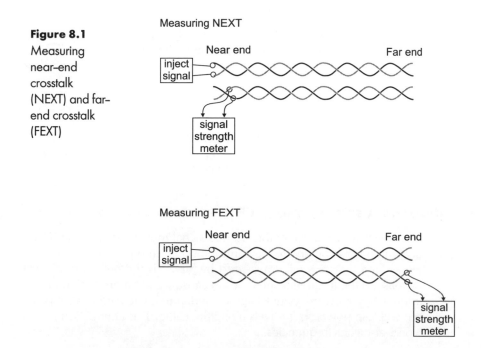

measured any signal found at the same end, called the 'near end', of another pair in the cable.

We check each pair and the specification refers to the worst result that we achieve. A typical spec is written as NEXT Worst Pair (dB) and may have a value of 66 dB at 1 MHz and 36 dB at 100 MHz. Remember, the bigger the number the better.

The figure also shows the way that we can inject a signal and check for the crosstalk at the far end of another pair. This is called far-end crosstalk or FEXT.

Equal level far-end crosstalk (ELFEXT)

Whereas the value of FEXT is measured, ELFEXT is a calculated value.

To calculate the value, we subtract the value of the cable attenuation from the FEXT value. The result is the ratio of the strength of the crosstalk compared to the received data signal, both measured at the far end. For reliable communications, it is vital that the disturbance at the far end due to crosstalk is less than the strength of the wanted signal, otherwise increased errors will occur at the receiver.

All pairs are checked and the quoted value in specifications is given as ELFEXT for the worst pair and is typically 64 dB at 1 MHz and decreasing down to 24 dB at 100 MHz.

Power Sums

When we took the measurements for NEXT and ELFEXT, we were applying a signal at one end of a pair and checking the crosstalk level on another pair either at the near end for NEXT or at the far end for FEXT.

In many real-life situations a pair will be receiving crosstalk from each of the other pairs in the cable rather than just the single one used when measuring values for NEXT and ELFEXT. We can reasonably expect the Power Sum values of PSNEXT and PSELFEXT to be a few decibels worse than the non Power Sum values of NEXT and ELFEXT.

Attenuation to crosstalk ratio (ACR)

If we want to use a cable, the wanted signal must be stronger than the unwanted noise otherwise we cannot distinguish one from the other. If we assume that crosstalk (measured by NEXT) is the only noise, it will increase as the frequency increases because we get more EMI at higher frequencies. Now the wanted signal is fed into the near end of the cable but will lose power as it travels along the cable. This attenuation is also worse at higher frequencies.

At some frequency the attenuation will reduce the signal level to exactly the same level as the crosstalk and at that point the signal and the unwanted noise have the same value. This is the upper limit to the frequency that we can use on this cable. This is not strictly true because there are some fancy dodges we can do to cancel some of the crosstalk and squeeze a bit more bandwidth but it is near enough.

Have a look at Figure 8.2 – this may be helpful. It's not easy but be careful to check that the graph is plotting signal strength against frequency. When we see a graph like this we tend to assume that the bottom line is distance along the cable but in this case it is frequency.

Figure 8.2
As frequencies increases, noise gets worse and the signal weaker

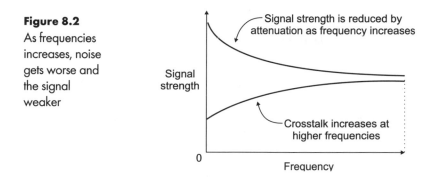

How fast does the signal move through a cable?

Light goes at 3×10^8 meters per second or 186 000 miles per second but signals on cables don't go that fast.

We may remember from Figure 3.7 that a cable includes capacitance and inductance. Both of these contribute to slowing down the flow of current along a cable. A voltage cannot move until the next bit of capacitance has had time to charge and inductance opposes any build-up of the charging current because the changing magnetic field creates an opposing voltage.

The more inductance and capacitance there is, the slower the signal can move and is governed by the formula: velocity $= \dfrac{1}{\sqrt{LC}}$ meters per second or miles per second depending on the units used for the values of inductance (L) and capacitance (C).

In most cables, the velocity is between 60% and 90% of the speed of light and in catalogs and data books is listed as 'velocity ratio' or 'nominal velocity of propagation' (NVP), so a velocity ratio shown as either 0.7 or 70% just indicates that the transmission speed is 0.7 times the speed of light. The symbol for the speed of light is c so a typical transmission speed would be quoted as 0.7c.

Delay skew

If we have more than one pair in a cable there is always a slight difference in the time that it takes a signal to travel along different pairs in the cable. This is partly due to manufacturing tolerances and also to the deliberate change in the rate of twist applied to each pair to reduce crosstalk. If we twist one pair more tightly than another, it will take up more wire and this pair will be longer than another more loosely twisted pair and so the total length of the pairs will differ slightly. The difference in the time taken for a signal to pass along the fastest compared with the slowest pair is called the delay skew and is measured in nanoseconds $(1 \text{ ns} = 1 \times 10^{-9} \text{ s})$.

If we are using each pair to carry completely separate signals, it causes no problems but in some protocols two pairs carry part of the same transmission to save time. If a serious delay occurs, parts of the same signal arrive at the far end out of order and the overall signal is scrambled.

Protecting cables by design

Sheath types

As we saw in the last chapter, the copper cores of a cable are each coated in a plastic insulation and surrounded by a sheath that may contain a metallic layer for screening or for EMI protection, or both.

There are usually two or three layers and the outer one is usually polyethylene to replace earlier cables that were often polyvinyl chloride (PVC). The big drawback with PVC is that in a fire it gives off dioxin, which is highly poisonous and in many cases more dangerous than the fire. PVC is not as waterproof as polyethylene. Safer in a fire is low smoke halogen-free fire retardant (LSHF-FR) cable.

The sheath types and their main components shown starting from the center are listed in Table 8.1. Most sheaths are a combination of aluminum, steel and polyethylene. In some cases, we see the word 'bonded' before the sheath type. This indicates that the outer plastic layer is bonded or glued to the underlying metal.

Table 8.1 Sheath types

Sheath	Main materials (from center outwards)
Alpeth	Aluminum–PE
Reinforced Alpeth	Aluminum–PE–steel–PE
Alvyn	Aluminum–PVC
ARPAP	Resin–coated aluminum–PE–aluminum–PE
ASP and Stalpeth	Aluminum–steel–PE
Steampeth	Aluminum–steel–PE–polybutylene (polybutene)
Plain lead	Lead
Polyjacketed lead	Lead–PE
PAP	PE–aluminum–PE
PASP	PE–aluminum–steel–PE

PE = polyethylene, PVC = polyvinylchloride

Outer protection

Direct-buried cables have a hard life. They may live in wet conditions, they may be mechanically damaged during installation, be suspended in overhead cables or attacked by rodents.

For any direct-buried cable we prevent water penetration by filling all cavities with a waterproof gel. Cables treated in this way are referred to as 'filled'. They also have additional protection as suggested by the environment. All parts of additional armoring are also filled with a material to provide waterproofing and corrosion resistance.

Armored cable that is underwater

In (relatively) gentle conditions like crossing a small stream, light wire armor is enough. This consists of a layer of jute, covered by light galvanized wire and then an outer layer of jute, which is a strong natural fiber used for sacking and rope making.

When our cable crosses lakes or coastal waterways we beef up the protection by using single wire armor – this is the same jute, wire, jute pattern but we use heavier wire.

If conditions get troublesome with strong currents and rocky bottoms we just double up on the wire layer. This gives two layers of steel wire separated by jute. The sequence is jute, wire, jute, wire, jute.

Mechanical protection

In situations where a cable is pulled through holes bored underground it is very likely that abrasion is going to be a problem and we protect the cable with a steel layer covered by polyethylene. The steel is often corrugated to provide some flexibility.

Aerial tape armor

In this case, a double layer of steel tape is wound around the cable. From the center out, the layers are jute, tape, tape and polyethylene. By using steel tape, we get mechanical protection and low frequency EMI shielding.

Gophers, rats and other rodents

Rodents attack underground cables but not just to annoy us. They have a problem with their teeth. Their teeth grow and, if ignored, will become so long that they will not be able to feed and can starve while surrounded by food. The solution, from their point of view, is to chew on something tasty like a nice cable. Rodents can be discouraged by applying an outer layer of steel tape to buried cables.

Protecting cables by conduit and cable trays

Instead of using heavy outdoor reinforced cables, we can protect interior cables to prevent abrasion and relieve stress by offering support to horizontal cables.

If, for example, we are passing cable through a wall, we can feed them through a metal or plastic tube, which is called conduit. When the cable is running horizontally, we need to support it at intervals of about 150 mm (6 inches) to prevent stress due to the weight of the cable. These horizontal supports may be a flat tray or a series of horizontal struts called a ladder rack, or ladder tray, as shown in Figure 8.3. Adding a lid enables the cable to be fully enclosed and it is then referred to as a raceway.

Figure 8.3
Cable trays and ladder racks to support horizontal cable

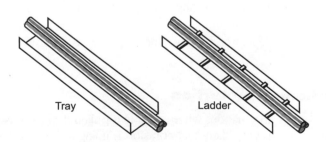

Tray   Ladder

Electrical protection

If an electrical circuit is happily sending data and then suddenly stops, there are two likely causes: a voltage surge or a supply failure.

Voltage surge

A voltage surge or 'spike' is likely to be caused by a spike on the electrical supply as a result of a switching problem with the utility company or a nearby lightning strike. These voltage surges can be of very short duration and occur quite often. They are too fast for us to notice but can do much damage to data-handling equipment.

A typical surge protector does not cost very much to buy and installation is just a matter of connecting it to the incoming mains supply but its performance is impressive. It can handle a pulse of 39 000 amps (A) and protect against EMI and will provide a safe output supply to equipment and telephone lines. A typical office or home product is made by a company called Belkin and they are confident enough to offer a lifetime replacement guarantee for any equipment killed by voltage spikes.

Supply failure

Just like breaking a leg, there is no convenient time to suffer a power failure.

If we happen to be transferring data at the moment of the outage then, to prevent loss, the speed of restoring the power is the main factor. If we have to send someone along to switch on the emergency generator, then the data is lost. In fact, by the time we notice the power failure it is already too late so we must have an automatic system – and it must be a fast one.

Uninterruptible power supply (UPS)

All UPSs use a battery or bank of batteries as the backup power supply. The art of the design is to be able to change from the mains supply to the battery power before our equipment loses its data.

Before installing a UPS it is worth investigating what the supplier means by 'uninterruptible'. Some systems called uninterruptible actually interrupt the supply for a short period. It does this by first detecting the loss of power and then switching the battery on line. So, how long does the switching take to occur?

During this time, the output voltage is falling rapidly as our equipment gobbles up the remaining power. We can prolong the power a little by adding capacitance which acts as a temporary store for some more electricity. This system hangs on a thread – will the new supply kick in before the old one dies away?

85

A real uninterruptible system is shown in Figure 8.4. In this equipment, there is no delay at all so our equipment does not notice any change of power level. In this circuit, the input alternating supplies are rectified to provide a direct current (DC) voltage used to charge a battery. The output DC voltage from the battery is then passed through an inverter that changes the DC back to alternating current (AC) to supply all our equipment.

Figure 8.4
Uninterruptible
power supply

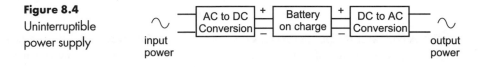

If the input power fails, the output continues to be supplied with no interruption at all. How long this backup can continue is only limited by the storage capacity of the battery and the power drawn by our equipment. The battery can be a single battery or a large bank of batteries, depending on the length of supply failure that we wish to guard against.

Connecting cables

When choosing a method of connecting cables we must be careful to check that it is approved by appropriate regulations and follows the recommended practices for the section of the industry involved. A method that is perfectly acceptable for, say, building a piece of electronic equipment may well be a disaster if used in a marine environment.

Insulation displacement connector (IDC)

This is one of the most popular methods used in telecommunication wiring in buildings and also to make connections inside equipment like computers. It is fast and easy, reliable and doesn't need much preparation before we start.

If we take a plastic-covered wire and cut into it with a knife we will go through the insulation and hit the copper core. At this stage the knife blade makes electrical connect with the wire. This is the point when we remember that we should always cut the power off when cutting into cables.

An IDC connector works by exactly the same method. It has two blades as shown in Figure 8.5. All we have to do is to push the wire into the gap between the blades – a process that we call 'punch down'. The shape of the blades must be adjusted to accommodate solid or stranded cores, though there are 'universal' types that accept either.

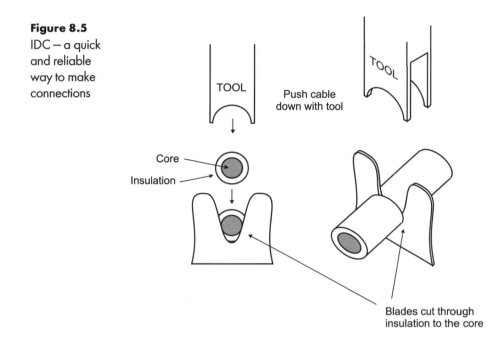

Figure 8.5
IDC – a quick and reliable way to make connections

TOOL

Push cable down with tool

Core

Insulation

Blades cut through insulation to the core

The IDC connector is found in different generic forms such as 66 block, which tends to be used for voice circuits, and the newer design called the 110, which is used mainly for data circuits, though the voice/data division is not total. There are also propriety designs such as QDF and Krone.

Tools come in three flavors. The basic design is simple and looks rather like a small screwdriver with an end shaped as shown in Figure 8.5. More sophisticated is the spring-loaded version. We push down until the pressure reaches the point where an internal spring is released and slams the wire down with the optimum pressure. The final option is to cut off the wire to the correct length at the same time. With a little practice, this makes for fast, reliable work.

Exothermic bonding

This can be used to weld cable to cable and for cable to ground rod to provide system grounding. Many cables are, of course, copper but this method can be used to weld a range of metals such as stainless steel, cast iron, common steel, brass, bronze and Monel.

The method is marketed under a variety of names such as Cadweld, Techweld and Thermoweld. The general appearance of the equipment is shown in Figure 8.6.

Figure 8.6
Strong, quick
and long lasting

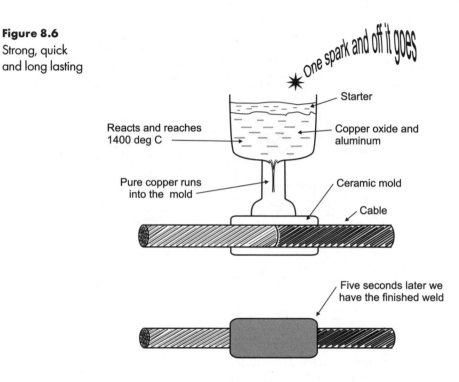

The starter is ignited very easily with a flint lighter and this starts the reaction between the copper oxide and the aluminum. The chemical action generates a great deal of heat, hence the name 'exothermic', and the result is pure liquid copper and aluminum oxide slag, which is discarded. The copper pours down into the ceramic mold and produces a molecular bond with the cable. The ceramic mold can be left in position or broken off when the weld has cooled.

The effect of the molecular bonding is that the copper and the cable become one single piece of material. The weld is permanent and will not loosen or suffer from increased resistance during the lifetime of the installation. It only takes a few seconds to do and the cost per weld is not frightening.

Clamps

These are purely mechanical methods. They are really just a steel tube holding the cables with bolts that can be tightened up to hold them firmly. In some cases the bolts have a ratchet action so that they cannot be loosened over their lifetime. As with all connections, we must check with the codes of practice to ensure the proposed design is acceptable. A saddle clamp is shown in Figure 8.7.

Figure 8.7
A saddle clamp

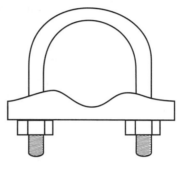

Soldering

This process is used for copper-to-copper connections usually within electronic equipment.

The advantage of soldering is that it can connect copper wires together or onto a connector both quickly and easily using only simple tools. The disadvantages are that corrosion sets in if the soldered joints are exposed to dampness over a period of time. The connections need to be protected either by being in a dry environment or by being covered by a suitable sealant. The other disadvantage is that poor workmanship is not always obvious but can cause weak, intermittent connections.

There are three vital ingredients – cleanliness, flux and sufficient heat.

Procedure

1. Gather the tools for the job. The source of heat must be sufficient to heat the two items to be soldered to about 200–250°C and for small wires an electric soldering iron of 25–60 watts (W) will be sufficient. Thicker cables will require something more impressive like a blow-torch. As for the solder, the most common type is a compound of 60% tin and 40% lead, though a lead-free version is available but requires a slightly higher temperature, around 300°C. Another higher

temperature solder is aluminum solder. This can joint a wider range of material such as aluminum, brass and stainless steel.

2. Clean both surfaces. We can use fine abrasive here. If stranded wire is tarnished it is almost impossible to clean so the best answer is to cut the cable back until we get to a clean section.

3. Either apply flux to the surface or use flux-cored solder. The flux prevents the clean surfaces from oxidizing as they are heated and encourages the molten solder to flow over the copper surface.

4. If using an electric soldering iron, apply the heat to the joint. If possible, it is better to hold the soldering iron below the joint so that any excess solder can be taken off with the iron. As soon as it starts to warm up, hold the flux-cored solder against the wire. The flux coats the area a moment before the solder melts. If all is well the solder will spread evenly through the joint. When the heat is removed the solder should set to a smooth shiny surface.

If using a blowtorch, we usually use flux and solder separately. Start by applying some flux, then start heating the joint, the flux will melt and protect the surface. Once it has reached the required temperature the solder will run through the joint wetting all the surfaces.

Soldering disasters

After completing a soldered connection have a careful look at it. Figure 8.8 shows one good soldered joint and two, possibly three, failures. Connection A has not been cleaned properly. The solder has not wetted the copper and has formed a globule of solder. Connection B has not been heated enough to make the solder really liquid. It is more like a paste and has a dull, rough surface. Attempt C is buried in solder and we have no way of knowing what is happening inside the joint. Sometimes this occurs when a doubtful joint is 'repaired' by adding more and more solder until a nice shiny surface is obtained. Good joint or bad – we have no idea.

Figure 8.8
Does it look
OK?

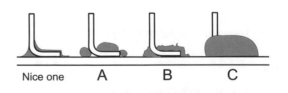

Nice one   A   B   C

The good joint is recognized by a shiny surface, the solder curving smoothly into the copper surfaces as the solder wets the surface with only just enough applied to enable the outline of the copper cable to be clearly seen.

To repair these disasters we must first remove the solder, clean the surfaces if necessary and start over. The best way to remove excess solder is by using a solder sucker, which is a spring-loaded hand tool that, when activated, sucks the solder off the joint. If not available we can use some copper braid from a spare length of coaxial cable or, as a last resort, clean the soldering iron and place under the connection. The solder will melt and pour onto the bit of the iron and can then be lifted off.

Chapter 8 quiz

1   As the signal frequency decreases, cable attenuation:

(a) increases.
(b) changes but may increase or decrease.
(c) decreases.
(d) remains at a constant value.

2   ELFEXT is equal to:

(a) FEXT – cable attenuation.
(b) FEXT + PSELFEXT.
(c) ELFEXT – ACR.
(d) FEXT + ACR.

3   A typical speed of transmission along a cable is:

(a) nearly 186 000 meters per second.
(b) 1.3c.
(c) 0.7c.
(d) 300 000 000 miles per hour.

4   A UPS always:

(a) includes a battery.
(b) interrupts the supply for a very brief moment.
(c) holds enough charge to close down the attached equipment.
(d) cannot be connected to an AC supply.

5   Exothermic bonding:

(a) corrodes seriously in damp conditions.
(b) requires the use of flux before use.
(c) makes the connection by melting ceramics into the cables.
(d) will not loosen over the lifetime of the installation.

Networks

As soon as we find some information there is always a need to share it with someone else. This need resulted in speech and writing and, later on, the telephone. Once we developed computers, fax machines and other things, we set about interconnecting them to exchange more and more information.

The interconnections are called networks and we have developed three sizes: big medium and small. The big one is called a wide area network or WAN. It covers very large areas, perhaps part of a country, a whole country, a continent or the whole world, rather like the Internet. Because of the large distances that are involved, it tends to use radio links, telephone lines and satellite links rather than the more local cables systems that we are interested in.

Just as a country is composed of a large number of cities, a WAN may be connected with many medium-size networks that generally cover areas up to cities or other areas within a country. This size of network is called a metropolitan area network or MAN. These MANs (not men!) are usually privately operated by large corporations, by local governments or by libraries.

The smallest network is called a local area network or LAN. It can be very small and technically it only needs a minimum of three addressable points or 'nodes'. This means that there are three things that we can communicate with, perhaps three computers, or two computers and a shared printer. A LAN can be much larger than this and there may be many LANs within a single building or perhaps just one. If we wish, we can connect LANs to other LANs, MANs or WANs.

LANs are a good idea

In fact, they are almost essential in all but the smallest organization.

They allow all of the staff or selected people to access files and data. The files are stored on a server, which is a computer with a large storage capability. This has the advantage that only one copy of a program needs to be bought and everyone can use it. A license is needed for the number of people using the program but it is still cheaper than buying individual copies. It also means that everyone is using the same version and so compatibility problems will not occur.

This compatibility issue occurs with the data as well. We only want one copy of a contract to be in existence otherwise it would be difficult to make any modifications without ending up with multiple versions of the same contract with all the legal problems that could cause.

Once we have a central store of files, we can devise a system to control access to it. We can organize automatic file backups and allow access to files on a need-to-know basis. For security, it is then possible to keep a record of the employees who have accessed a file together with the time and date or, at least, which computer was used.

Internal communications can be fast and easy by providing an internal e-mail system, which should reduce the time spent as people wander around the offices and, theoretically, reduce the amount of paper used. Unfortunately, the ideal of the paperless office is getting no closer and is probably receding.

Network architecture

In deciding on the design of a network and a LAN in particular, we have to consider four topics: communication medium, topology, data control methods and the speed of transmission required. It is these four factors that differentiate one design of network from another. We will have a brief look at these options now.

Medium

There is a range of transmission media used in networks from copper cables to optic fibers, radio links and infrared links. We will have a look at all their various merits and associated problems.

Copper cables for LANs

In Chapter 7, we met all the cables that we use in LANs, so for now it will be enough to concentrate on the use rather than delve back into their construction and operation.

Unshielded

The good news about unshielded twisted pair (UTP) cable is that it is relatively cheap to buy, does the job and the installation methods are well established although for best results we must adhere to the installation practices. What more could we possibly want? Well, there is the small problem of electromagnetic interference (EMI). Although the twisting of the pairs does help, it can't replace shielding and screening. The length of the links between nodes is limited to about 100 m or 328 ft.

Screened twisted pair (ScTP)

We get fewer problems with electromagnetic fields but this advantage can be lost by any break in the shield or incorrect grounding. The shield also makes the cable heavier and thicker and makes installation more difficult.

Shielded twisted pair (STP)

This cable, as you remember, has shielding around each pair as well as the overall shield as used in ScTP cable. As we would expect, this gives us much better crosstalk and EMI protection. However, the disadvantages are much the same as ScTP cable. Breaking the shield or improper grounding will cause problems and it is significantly heavier and bulkier to install.

Remember that ScTP and STP cables have different characteristics and are not interchangeable without careful thought.

Coaxial cable

These cables provide good shielding but there are many slightly different designs with a range of specifications and applications. They tend to be used for the final link to the equipment rather than for the complete LAN.

Fiber optics

These operate by passing light signals along a glass or plastic cable. They provide complete immunity from EMI and radio frequency interference (RFI) and they do not radiate any interference to surrounding equipment or cables. They are also able to transmit signals over long distances and at faster rates and provide much of the communications for WANs, telephone systems and transoceanic communication cables. On the other hand, they tend to be more expensive and more difficult to connect and install.

Since we often come across fiber-optic cable during installation, there is a brief introduction in Chapter 13 but to find out more about how it works and how to use it, there is our sister book *Introduction to Fiber Optics* [Crisp, 2001].

Wireless systems

In some situations in which a cable is not a convenient option, perhaps because of the terrain or if we have to cross land without having permission to lay a cable, it may be convenient to send the information by means of a radio or infrared link. We may also require part of a LAN to be a mobile link in a vehicle.

It is expensive and often does not match the performance of copper cable but high-speed links can be designed. We must think carefully whether there are any other options available.

Radio

Popular uses are in battlefields, hospitals and warehouses where mobility makes permanent links difficult.

The performance of a radio link depends on the frequency used. At most frequencies the radio link can provide long-distance transfer of data and is able to pass through buildings and other obstructions. The radio signal is broadcast in all directions, so the position of the receiving station is not critical, rather like a local broadcast radio station. If the data is at all sensitive, it is open to easy interception by anyone with suitable equipment. We can introduce some technology like scrambling the data or using frequency-agile radio systems. These radios are very smart. They change their operating frequency, jumping about the frequency spectrum so that the receiver must be specially coded to change its receiving frequency in the same sequence as the transmitter. A normal receiver would not receive anything useful at all.

As the frequency increases, the radio signal is easier to concentrate in a narrow beam just like the headlight on a truck. This means that by concentrating the beam we can reduce the input power for a given range but it also means that we must have a line-of-sight positioning of the antennas. This system using microwave radio transmissions is very popular with telephone and broadcasting organizations and their small parabolic dishes are to be seen on their towers scattered around the countryside.

The use of radio signals has a couple of possible drawbacks. The radio signals may pose some health risks though this is still a matter of conjecture but the widespread use of mobile phones will encourage the necessary research. At one time I was involved with the maintenance of radar systems and it was noticeable that every time the effects of their transmissions were investigated, it resulted in the allowed exposure being reduced.

Infrared

Infrared links can send data as we do with some television remote controls and just like these remotes we can sometimes bounce the energy to go around an obstruction. Many infrared links use infrared lasers that give longer range but the beam width is so narrow that alignment is very difficult during installation and has to be maintained during the life of the link. Infrared is also prone to interruption by fog, rain, snow or even birds flying through the link.

Topology

If we are to use cables to interconnect a building full of computers in a LAN it soon becomes apparent that the design of the network needs to be thought out carefully before we rush in with reels of cable.

The design of the network is called its topology and the layouts are described with the usual handful of technical names.

In looking at three or four topologies, it is worth remembering that a real network may well employ more than one type in the same layout and we also meet modifications of these basic designs. As with all choices we find that each solution is a combination of advantages and disadvantages, so we have to make a value judgment based on what is likely to be most important in the present situation. In a city office we may well go for speed and low cost whereas in a war zone reliability and security will be higher on our 'must haves' list.

Mesh topology – simple but seldom used

A simple topology, but one suited only to low numbers of nodes, is shown in Figure 9.1. The instructions to the installer would be really easy to understand though not so easy to carry out – 'just connect every device to every other device by a cable' – and the job's done.

As we have seen, there are good and bad points in every design. This one can work well enough for four or five nodes but becomes quickly troublesome as the number of devices, or nodes, increases. The first problem would become obvious when we calculate the total length of cable and installation time involved. The next will occur when we try to

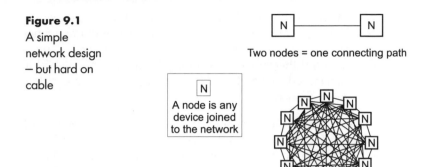

Figure 9.1
A simple
network design
— but hard on
cable

Two nodes = one connecting path

A node is any
device joined
to the network

Twelve nodes - how many connecting paths?

connect all the cables to each device. We could easily be faced with a 1000 cables to be connected to a printer.

For small networks, there are possible benefits to this design. Whichever device we want to send data to is connected by a direct link from our computer and the data will not be delayed through switching circuits and so our data will run at maximum speed. There is also a gain in reliability as data can be switched to alternative routes in the event of a system failure.

This is a bit like delivering data around a city by having a 100 trucks to make a 100 deliveries. If each truck delivered to a single address we would use a lot of trucks but at least it would be fast.

Hybrid mesh topology is a halfway stage in which only some of the more important nodes have single connections. This gives a degree of redundancy without enormous numbers of data connections.

Overall, this system is seldom used and the real solutions are based around one of the three topologies that follow.

Bus topology

This is a much simpler system. We have a single cable to which all devices are connected and all can use it as a common path to send data to any other device. When one of the devices wishes to send a signal to a single node, we just put the signal on the cable and it moves both ways along the cable, making contact with all attached nodes. Some 'address' data is included to ensure that the data is received by the correct device and ignored by all the others. This is just like sending our truck out with a series of addressed packages to be dropped off along a single route.

If we just left the ends of the cable disconnected, we have a danger of reflections occurring as the signals hit the sudden change of impedance at the open circuits. You may remember that we had a look at these reflections and cured the problem by adding impedances matched to the characteristic impedance of the cable. In a bus, we call these matching impedances 'terminators'.

If a terminator fails or we cut the cable anywhere along its length we will cause serious reflections and this will result in loss of communications to all of the computers and other equipment connected to the bus. Worse than that, it can be quite a job to track down the cause of the failure. Bus topology is not recommended for sending critical data along non-secure cables.

If you skipped all this stuff about impedance matching because it was too boring, or have forgotten it, then we can flick back to Chapter 3. We may find this bus topology described as 'linear bus' topology and is shown in Figure 9.2.

Figure 9.2
Linear bus

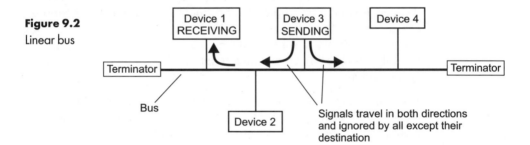

Star topology

More popular and easier to install than the other designs. This is the usual topology used in buildings except very large ones that may use a hierarchical ring as we will see in a minute. The star topology puts one device in the center of communications, a bit like a telephone exchange with all our houses connected to it. The central device is called a hub but it also goes by a variety of other names by different suppliers. We also hear it referred to, in certain circumstances, as a repeater, concentrator, MSAU, MAU (multistation or media access unit) or CSAU, CAU (controlled station or controlled access unit) or a switch. Whatever it is called, it sits at the center and a cable is taken to each node. All communications within the network pass through this hub.

This is simple to connect and has the advantage that a failure in any node will leave the rest of the network unaffected. The weak link is the hub. If this fails then we kill off all the communications between any of the remaining devices (see Figure 9.3).

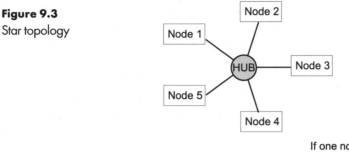

If one node fails the
others can still work

Hierarchical star topology

A hierarchical star is an extension of star topology and is used in very large buildings. As we can see in Figure 9.4, it is just a ring topology but instead of just having a node on the end, it has other hubs. It looks a little bit like a bunch of flowers. This allows a building to have an overall star topology with a separate hub on each floor – but more about this in Chapter 10.

Ring topology

If we look at Figure 9.1 again, we may notice that there is a single connection that goes around the outside from node to node. Each

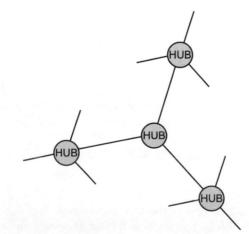

device has an input connector and an output connector. These connectors are generally called 'ports'. If the output port of each computer is connected to the input port of the next one, we can join up all the computers with the last one being joined to the first in a ring, as the name suggests.

In use, when a message is sent from one computer it leaves its output port and is passed around the ring in a predetermined direction. It has an address attached, so as it passes through each computer it is ignored by all except its destination.

Since all data passes around the loop, a failure of any part of the loop will stop it working. A double-ring system can provide some improvement in reliability by adding a spare loop so that it can be switched in if required.

Faultfinding is easy as we can follow a signal around the loop until it stops. The faulty device or link will have a signal input but no output but adding extra nodes to the loop can be more troublesome than with other systems like a star.

Tree topology

There is nothing new here except the way we have combined different topologies that we have already met. There are, of course, many combinations of the basic topologies just as we can make many words from combinations of the same letters. The tree topology is shown in Figure 9.5. In this diagram there are two boxes called 'bridges'. These are included to control the data flow between the bus and the stars – don't worry about bridges for the moment we will come back to them a bit later.

Figure 9.5
Bus + rings =
tree

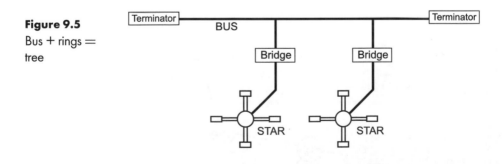

The advantage of the tree topology is that the attached segments can enjoy the benefits of several star systems supplied by data from a single bus. Ethernet systems often use tree topology but having several topologies being fed from a single bus can result in serious disruption if the bus fails.

Physical and logical networks

Sometimes we can look at a network and the appearance can be misleading. It may appear to be one type and yet behave like another.

What it looks like is called the physical network and what it behaves like is called a logical network and strangely they can be different, so what we have to do is to ask the data what sort of network it is in.

If we look at Figure 9.6 we would say straightaway that is a star network so it is a physical star but what about the data's point of view? The data would think it was a ring network because, inside the central point where we can't see it, the cables connected from each node are actually connected across to the cable that leads to the next node. The data is therefore going from node to node rather than being fed from the central hub. Figure 9.7 shows the real data movement and why it is a logical ring network. This is often referred to as a star-wired ring or collapsed ring.

Figure 9.6
What type of network is this?

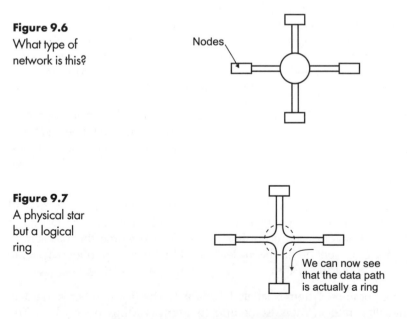

Figure 9.7
A physical star but a logical ring

Figure 9.8 shows a physical star but a logical bus. The bus has been reduced in length and so the nodes connected to the bus appear to radiate out like a star topology but the nodes are not fed by a central hub but simply take data off of the bus. This layout is called a star wired bus.

Figure 9.8
It looks like a
star but it is
really a bus

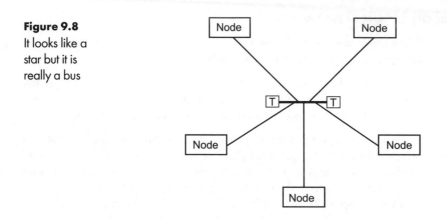

Ethernet

This is a LAN technology first developed in 1972 and now accounts for something like 80% of all LAN networks. It allows the use of twisted copper and coax cables of various types together with optic fibers and supports a wide range of data speeds. In 1972 the operating speed was 1 Mb/s and it is now 1 Gb/s – a 1000-fold increase in 30 years. Now there's an interesting thought for our grandchildren.

It aims to provide a simple, low-cost system with a high degree of compatibility and one that can be easily expanded. It has been adopted as an international standard by the ISO. It was designed as a bus topology but more recently it has become a logical bus similar to that shown in Figure 9.8.

As the Ethernet title includes such a wide range of data speeds and cables it is never enough just to say that we have installed an Ethernet system without further details. To overcome this problem, a naming convention has been developed that will help to identify some details about the system.

One common system is labeled 10Base-T. The first number is the data signaling rate in Mb/s (sometimes written as Mbps or Mb s^{-1}). The standards at the moment include first numbers of 1, 10, 100 or 1000.

The second part indicates the transmission method used and will be 'Base' or 'Broad'. Baseband transmission is where the whole of the bandwidth is used to send a single analog or digital signal at its original frequency. This cuts down on the electronics but only allows a single signal to be transmitted at one time. Broadband allows several different signals to be transmitted at the same time through a single cable (see Chapter 5). This makes better use of the cable. We have combined and separated just two signals but we can do equally well with hundreds or thousands. Telephone companies do this in a big way.

The third part of the Ethernet name is either a number, which is the maximum segment length of the network, or the type of media used. Some of those in use at the moment are listed in Table 9.1.

Controlling the flow of data

If we have, say, a ring topology connecting several computers and printers, all of the data is passing through each of the nodes. When the data is confidential we don't want it to be downloaded to each of the computers. Even if it is not a matter of secrecy it would still be very inconvenient if we clicked on 'print' and all of the attached printers sprang into life.

To overcome these problems we need to introduce a form of control that would perform the function of an address. There are different methods adopted depending on the technology that we are using.

The control of data on a ring topology is a good starting point.

Table 9.1 Types of Ethernet

Ethernet spec.	Bit rate	Cable type	Segment length
10BASE-5	10 Mb/s	50 Ω thick coax	500 m (1640 ft)
10BASE-2	10 Mb/s	50 Ω thin coax	185 m (607 ft)
10BROAD-36	10 Mb/s	75 Ω broadband coax	3600 m (2.25 miles)
10BASE-T	10 Mb/s	Two-pair Cat 3 UTP or higher	100 m (328 ft)
10BASE-F	10 Mb/s	Two multimode optic fibers	2000 m (1.2 miles)
100BASE-TX	100 Mb/s	Two-twisted pair UTP Cat 5 or higher or STP	100 m (328 ft)
100BASE-T4	100 Mb/s	Four-pair UTP Cat 3 or higher	100 m (328 ft)
100BASE-FX	100 Mb/s	Two multimode optic fibers	400 m (1312 ft)
100BASE-T2	100 Mb/s	Two-pair Cat 3 UTP or higher	100 m (328 ft)
1000BASE-SX	1000 Mb/s*	Two multimode optic fibers	
1000BASE-LX	1000 Mb/s*	Two singlemode or multimode optic fibers	
1000BASE-CX	1000 Mb/s*	150 Ω balanced shielded copper. This is a special cable – not standard STP	25 m (82 ft)
1000BASE-T	1000 Mb/s*	Four-pair UTP Cat 5 or higher	100 m (328 ft)

*1000 Mb/s = 1 Gb/s.

Token ring

Let's consider a similar situation in real life – not that topology is not real life but you know what I mean. There are six truck drivers and one truck to which they are all allowed access – how do we control the use of the truck so that everyone doesn't try using it at the same time? One answer is to have one key that is passed from driver to driver. Whoever holds the key has the use of the truck until the job is finished and the key is passed to the next driver.

The token ring runs along these lines. If there is no information to be shared, a token is passed from node to node around the ring. This token is an electronic code that gives the node holding it the sole rights to use the ring. If this node has nothing waiting to be sent it simply passes it on to the next one. On a quiet day the token just flies round and round the loop.

Now, Figure 9.9 shows the case where something, perhaps a computer, called node 3 has information for another device at node 2.

Figure 9.9
Token ring

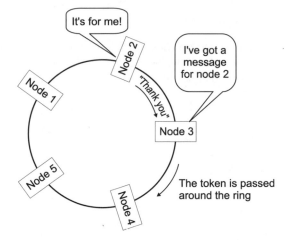

This is the sequence of events:

1. Node 3 waits until it receives the electronic token. It then attaches an address and the data to be sent.
2. The token, address and message is passed on to node 4 – it moves in a set direction. Node 4 reads the address, doesn't recognize it and so passes it on to node 5. This node does exactly the same and passes it to node 1. Then node 1 passes it to node 2.
3. Node 2 receives the data, recognizes the address and downloads the attached message.

4. Node 2 has one last job to do. It attaches an acknowledgement to the token and sends it back to node 3.
5. In this example node 3 just happens to be the next node so it receives the acknowledgement very quickly. If this was not the case, each node would do the 'check address and pass it on' procedure until it reached the sender.
6. Node 3 has one last job. It removes all the information attached to the token and then passes to node 4 and so on around the ring until another node wants to send something.

The token ring appears laborious but in reality we must remember that the token is traveling at about three-quarters of the speed of light and even allowing for switching and checking times is likely to go round the ring many thousands of times a second.

If these speeds are still too slow for us, we have the slotted ring as an alternative.

Slotted ring

This is a close relative to the token ring but there is a slight difference in the way that it works. Imagine for a moment that there is a group of 30 people and they want to be transported to the other side of the city. They could all keep together by waiting for a bus or we could split them into little groups of four and use a whole line of cabs.

This is the idea behind the slotted ring. Instead of waiting for a token to arrive and be able to attach the whole message to it, we can have many 'slots' that travel around the ring but these can only accept a fixed size package of data called a subunit, so a long message may have to be split up into small packages sent separately and reassembled at their destination. Smallish messages that can fit into a single slot can be picked up and transported very quickly.

Here is the procedure:

1. The sending node splits up the message into slot-sized subunits. To each piece it attaches an address and a note to say whether the message has now been completed, otherwise the receiving station will never know when the message is complete.
2. Each subunit is loaded into the next available subunit and sent on its way from node to node around the ring. As with the token ring, each node checks the address and if not recognized the slot is passed to the next node.
3. When the slot arrives at its destination, the subunit is stored and the slot returned to the sender where the message and address details are removed and the slot is then put back onto the ring and is available for use by other nodes.

4. At the receiving node, the whole message Is collected up a subunit at a time and the process is complete.

Demand priority

This is used on star topologies and the central hub acts as the controller.

As it is a star topology, all the data passes through the central hub and does not pass through the nodes unless they are to receive or originate the communication.

The nodes are connected to ports on the hub. The word 'port' is just a fancy electronic or computing word for a connection. These ports are each given a priority level during the setup of the hub. This happens much the same in a normal PC where we have ports that handle information from the floppy disk drive, CD drives, serial and parallel inputs. If we insert a floppy disk and a CD ROM and then switch the computer on it will receive information from both disks at the same time – so which does it read? The usual answer is the floppy disk because, during the original setup, we have put the input ports in an order of priority and have given the floppy disk drive port a higher priority than the CD ROM port but that was the installer's choice and it can be changed.

Have a glance at Figure 9.10. We will assume that node 3 has data for node 5 and, at the same time, node 2 has information to send to node 4.

Figure 9.10
Exchanging
data using a
star system

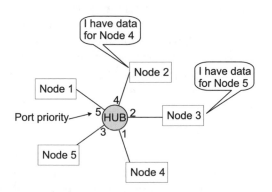

Here is the sequence of events:

1. The hub receives requests to transmit from nodes 3 and 4 but which message is dealt with first?
2. The hub checks for priority requests. The nodes can request one of two priorities – high or normal. In this case, both have the same priority so no decision can be made. The hub now takes its input

from the port that has been given the highest priority just as in the computer example. In the figure, the order of priority given to the ports are 4, 3, 5, 2, 1 so it will check for an input at node 4 but there isn't one, so it will then check node 3 and it finds a request waiting.

3. The hub tells node 3 to send its message. The hub then checks its destination and sends it to node 5. No other node is involved with this transfer.

4. Now there is only one node waiting to send, so the hub gives node 2 permission to send, and its data is then delivered to the destination nodded and the hub goes to sleep until another request is heard.

Carrier sense multiple access/collision detection (CSMA/CD) as used on the Ethernet system

It is a common experience in everyday conversations that a couple of people will sit in silence for a while and then both suddenly start talking at the same moment, then they both stop and then try again. Sometimes they will clash again but usually one waits for a little longer than the other and the conversation continues happily. This is basically how the CSMA/CD Ethernet system works.

The Ethernet treats all attached devices equally, there are no systems of priority, and so it often happens that two or more devices will try to send data at the same time. This is referred to as a data collision. Despite the emotive language, this is an ordinary event and quite expected – there is no data lost or damage done. It is just like getting a busy tone on a telephone line.

Ethernet frames and addresses

To send information it must first be organized into a recognized format called a 'frame'. A frame is a string of data made up of blocks called 'fields' each of a specified size and content.

The first two fields carry the two 48-bit (binary) addresses that hold the destination and the source of the message. The next field carries the data to be transmitted and the last field is a binary code which is able (usually) to detect whether an error has occurred during the transmission.

As the first 48-bit address is put onto the bus, all connected devices read this address to see if it is addressed to them. If it is not, they stop reading the information.

Sending information over the bus

Let's consider an Ethernet system on a single bus with several computers attached to it. There are no messages to be sent and all the nodes on the system continually monitor the LAN, listening for their own address, which would indicate that another node is trying to contact them.

In Figure 9.11, node 1 wishes to send some information to node 3 some distance away along the bus.

Figure 9.11
Collisions on the bus

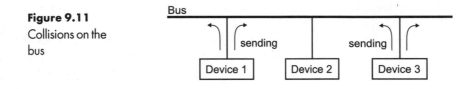

1. Node 1 is monitoring the bus and detects that someone else is using it by sensing the presence of a signal called a carrier. In this situation, it makes no attempt to send a signal. Once the carrier stops it can go ahead but so can anyone else.
2. A moment after node 1 has started to transmit, it detects node 2 starting to send. It now sends a jamming signal which effectively says 'wait a while and try again' and both node 1 and node 2 stop sending.
3. Both nodes now wait for a random period of time, just a few microseconds, and try again.
4. If a collision occurs again, then lengths of the random delay increase over a greater range and the one with the shortest random delay tries again. Unless the network is seriously overstressed this sorts out the problem. If collisions continue to occur for 16 consecutive attempts, the sending device gives up and abandons the attempt and we get a 'fail' notice on our screen.

Network connecting devices

No sooner have we built a network than someone says 'Yes, very good, just right, perfect but it would be better if it could just connect with ...'.

We may want to extend the network by connecting it with another network that may or may not be using the same technology, we may just want it to connect to a computer that is physically some distance away and beyond the normal operating range. These problems have given rise to a range of devices that can be employed as necessary. We will look at some of them, starting with the simplest.

Repeater

These are not very clever. They are basically just an amplifier just like the one that we can use at home to connect our little CD player to our impressive speaker system to irritate our neighbors.

They are extensively used in telephone systems. After all, if we want to talk to someone in the next town it would be unreasonable for the phone company to expect us to shout loud enough to reach there without amplification. Telephone lines therefore have amplifiers at regular intervals all along the route.

In a LAN it allows us to increase the physical length of a network connection, though we should be wary about using the repeater to add more and more nodes.

They often provide two small services in addition to amplification.

As digital signals pass along a cable, they accumulate noise and other losses have the effect of reducing the amplitude of the signal and rounding off the shape of the pulse. The repeater can be used to clean up the digital pulses, as shown in Figure 9.12 and hence give the pulses a fresh start.

Figure 9.12
Two benefits of
a repeater

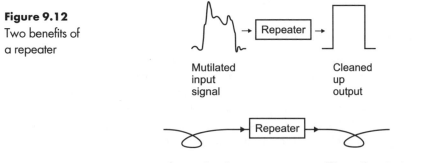

The other benefit is that the repeater can be used to interconnect two different media. It may be that the copper cable would like to pass its information over to an optic fiber. Since copper is an electrically based system and optic fiber runs on light, the repeater can provide the voltage/optic changeover as required.

Hub

This is very similar to the repeater. It includes an amplifier but it has two or more outputs, so it is a form of multiport repeater. Just like the repeater, it does not analyze the data, it just passes it on to the connected outputs, which may include other hubs to expand the network.

Bridges

You may remember that we mentioned bridges when we looked at tree topology in Figure 9.5. They are used when we want to join separate

LANs or segments. They can filter traffic and handle data collisions and other problems with the network. There are a few names that we will come across.

Medium access control (MAC) bridges

These are able to communicate between LANs of the same type like two token rings and can provide some limited services like changing the data rate to accommodate differences due to the input and output medium types. It allows 16 Mb/s STP to exchange data with a 4 Mb/s UTP system. This MAC has nothing to do with Macintosh computers.

Transparent bridging

These are found in Ethernet networks in which the incoming frames are read and forwarded to the correct segment. So, in effect, it acts rather like a mail system or telephone exchange.

Translating bridges or link bridges

These are one step up the evolutionary scale as they are able to connect network segments that use different access technology and can convert one frame format to another. Typical uses are passing data from Ethernet to token ring or Ethernet to fiber data distributive network (FDDI). FDDI is a fiber-based network using token passing on a double-ring optic fiber system running at 100 Mb/s.

Switches

A port can be connected to a variety of devices such as a server, a hub, a shared device like a printer or it can be connected to another switch. Switches behave like multiport bridges in which each port is connected to a separate network communication channel. They can be connected into an existing LAN to improve overall performance by creating a series of individual LANs, each with a dedicated communication channel.

Routers

These are also connecting devices that exchange data with a range of LANs and are rather similar to bridges and switches, but routers are more intelligent. The router is not only able to switch routes but has the ability to make a choice depending on such criteria as cost, availability of the route and the amount of traffic using a particular route. To help it with these decisions, it stores and updates a map of the whole of the available network including pathway utilization and availability from the router to every destination.

Chapter 9 quiz

1   A network that connects a maximum of 10 computers is most likely to be:

(a) MAN.
(b) WAN.
(c) PAN.
(d) LAN.

2   An Ethernet frame is composed of:

(a) rings.
(b) segments.
(c) loops.
(d) fields.

3   The purpose of the token in a token ring system is to:

(a) prevent more than one device from sending data at the same time.
(b) increase the speed of data transfer.
(c) allow data to be sent to more than one destination.
(d) provide a link to another LAN.

4   An Ethernet code of 1000Base-T would indicate that the:

(a) cable used should be at least Cat 3.
(b) maximum transmission rate is 1 Gb/s.
(c) medium should be (T)hinnet coaxial cable.
(d) minimum transmission rate is 1000 Mb/s.

5   If a data collision occurs in an Ethernet protocol:

(a) both data streams are corrupted.
(b) the first sending device to notice the collision will switch off and allow the other to continue.
(c) no data is lost.
(d) the hub will select a sender at random.

10

Cables in buildings and between buildings

Cables technologies do not last as long as buildings. For this reason, it must be assumed that the cables will, sooner or later, have to be upgraded or replaced or extended. In years to come we will be using technology that has not yet been invented so predicting the necessary cable is impossible.

When a building is being designed, we must include pathways that are actually the containers for the cables or their access routes. It is important that telecommunication routes or pathways are carefully thought out and included in the initial design and costing, as well as providing scope for later modifications. In the long run this will have a significant effect on the cost of installations and modification and repairs over the lifetime of the building.

The cabling for the building starts at the demarcation point (DP). This is the point where the responsibility passes from the Access provider to the owner of the building.

Whether we are involved with the design or the installation, we must take all steps to ensure the safety of the end users and our coworkers and to comply with all local regulations and applicable standards.

In this chapter we will look at cabling of multistory buildings. To break the task up into bite-sized portions, we will start from our desk where we have our computer ready to send some data. This will involve the wiring of the whole floor including individual offices and large open offices as are popular with current designs to provide maximum flexibility over the lifetime of the buildings. This group of cable layouts is called horizontal cabling.

Horizontal cabling

First and most important is to clear up a likely misunderstanding. The name 'horizontal' does not imply that the cable itself is actually horizontal. It may or may not be, it could be vertical just as easily. This is a bit like being told that all the departmental heads in a company were on the same 'level' – this would imply equal seniority and not that they were of equal height, sitting on the same size chairs or have their offices on the same floor of the building.

Horizontal cabling includes all the cabling between the telecom outlet (TO) in the work area and the cable connections in the telecommunication room (TR). We will meet the telecom room in a moment.

Pathways

Pathways include all the structures that keep the cabling safe by supporting and protecting it and must also allow easy and clear access to the cables.

Things to consider at the design stage

Maintenance and modification of the system

Building-in easy access has significant effects on the overall operation and cost of an installation. With the best will in the world, we will not do a very good job if we cannot reach the connections easily or we are trying to locate a fault without being able to see what we are doing.

The pathways, and indeed all installations, are best designed to accept standards-based components so that we are not tied to a single vendor for replacement materials. If we were to install raceways that could just about handle a specially designed highly flexible cable and then the supplier went out of business, we would have to revert to 'standard' cable which may not fit in the pathways. This could turn a quick, easy job into a real nightmare – and expensive too.

We should also ensure that we can gain access with a minimum amount of disruption to the work areas. There are some very unpopular 'what ifs'. What if the building's electrical supplies had to be closed down to access some cables? What if all the furniture has to be moved?

113

The only way to be safe is to ensure that the design and construction of all pathways must meet or exceed all relevant national and local codes and standards.

Although a single work area only requires two cable runs, it is a good idea to ensure that there is space and access to allow for at least three. It costs little at the installation time but could be expensive as a later modification.

Electromagnetic interference (EMI)

We should avoid running pathways near to potential sources of EMI. Such sources are: electrical power wiring, transformers, electric motors and generators, transmitters, induction heaters and arc welders. In Chapter 12 we will consider grounding, bonding and lightning protection.

Where are the pathways likely to be found?

In the ceiling

There is usually a space between what appears to be the ceiling when viewed from the office below and the real 'structural' ceiling which is part of the floor above – or the roof if we are on the top floor.

The office ceiling is suspended from the real ceiling and can only be used for horizontal cables serving the floor below. The ceiling has to be fully accessible from the room below and not too high. Height is generally less than 3.4 m (11 ft).

We may have local regulations restricting the use of this space due to its use for ventilation and air conditioning and limiting what can be installed. Usually, consolidation points are allowed but not other connectors or telecom equipment.

In large ceilings, the area is divided into zones of between 34 and 84 m^2 (365–900 ft^2) and pathways serve each zone rather as if it is a separate office.

Under the floor

Just as the suspended ceiling leaves a gap for the cables, the same trick can be played with floors. These are called access floors or raised floors. As with the ceilings, the gap may also be used for ventilation purposes. The floor is supported by pedestals and braces called stringers which support a paneled floor which is finished in carpet or whatever is required.

As with the suspended ceilings, it is not recommended that telecom equipment or connections other than a consolidation point are installed below the floor.

The pathways under the floors should be spaced sufficiently to allow us to stand on the structural floor without damaging the cables.

When panels are lifted for access, we must take suitable steps to prevent workers from falling down the hole. This would make us unpopular with everyone apart from the legal profession.

Or even in the floor

These floors are called cellular floor systems and serve as both a structural floor and as an in-built system of raceways or cells. There are also ducts that provide access to the cells. The floors can be fabricated from steel or concrete though, of the two, the steel is the preferred type. A section of a cellular floor is shown in Figure 10.1. Placed above the cells can be other ducts called trench ducts that run across the cells at right angles and allow access down into the cells. This provides easy right angle turns for the cable. Trench ducts have plenty of space for cables and, like the cells, provide separation for telecom cables and power cables. Trench ducts are difficult to live with as their covers tend to behave like the loose lid that they are. They feel and sound different and do not always remain level to blend in with the surrounding floor. They are not really a very good option and may be hidden underneath a carpet which can make access difficult.

Figure 10.1
The cables may
be in the floor

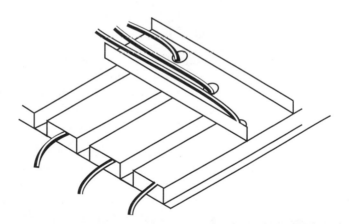

It may be in conduit

Conduit is really just a circular cross-section raceway and is available as a plastic or metal rigid pipe or in a flexible form. The cable is installed by pulling between pull boxes (PBs) – an example is shown in Figure 10.2.

Figure 10.2
Pull boxes

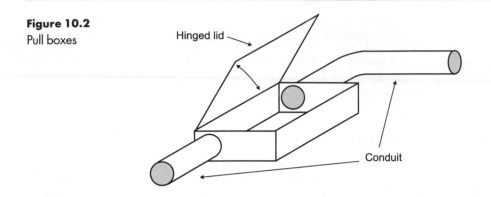

Hinged lid

Conduit

Access to the cable is by means of the hinged lid. To make it easy to pass the cable through a pull box, it makes good sense to choose one where the outlet is physically opposite the inlet so the cable can pass smoothly through. We don't want the cable getting lost inside or dragging its way around sharp corners.

Be careful not to damage the cable by applying excessive force. With four-pair cables, don't exceed 110 newton (N) (25 pound force).

When pulling cable, it is important not to fill the conduit too much as this reduces the possibility of adding more cable at a later date. Generally, restrict the filling to about 40% of bore for the first installation and thereafter we have another 20% still available. We should not fill the conduit over 60% of its capacity.

When the conduit is installed there is a small nylon cord passed through it called a drawstring, pull cord or pull string. This is attached to the cable to pull it through any of the pathways. Even so, it is not always easy but there is a pulling lubricant available to enable the cable to slip past the other cables and around the bends more easily.

To feed a pull string through a conduit we can use an air bottle to provide compressed air to blow a foam ball with a string attached.

The work area (WA)

Single offices

This is the office where we sit at our computer. Attached to the computer is a cable that we call a 'patch cable' or 'patch cord'. These cords are stranded flexible cables and are generally outside the scope of the regulations yet have a significant effect on the overall operation of the system.

To connect to the local area network (LAN) we have to plug the patch cable into a socket called a telecommunications outlet (TO), as shown in Figure 10.3. This outlet may be a wall socket, a surface-mounted socket or a floor mounted socket. They should be positioned so that the patch cable or cord does not need to be any more than 5 m (16 ft) in length and since it is highly unlikely that we will need a data feed to any device that is not electrically powered, it makes good sense for a power socket to be within 1 m (3 ft) of the outlet and at about the same height.

Figure 10.3
A patch lead connects to a wall socket

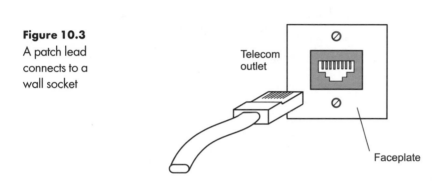

Telecom outlet

Faceplate

The wall-mounted socket should be between heights of 375 mm (15 inches) and 1220 mm (48 inches) and not back-to-back with a similar socket in an adjacent room. This could result in a significant reduction in the thickness of the wall at that point. This is not a structural problem – the building won't collapse but it could affect sound insulation.

The cable assignments shown in Figure 10.4 are compatible with all 100 ohm (Ω) twisted-pair cable but we may have to be careful with the patch cables which may be non-standard.

Have a look at the layout of the room to ensure that the position of a floor outlet and the associated cable will not cause people to trip.

Multi-user offices

For maximum flexibility many buildings are designed with vertical pillars supporting a series of open floors. In this way, the open space can be reorganized to suit a range of uses by temporary partitions and the use of furniture designed for this purpose.

In these large offices, it is impractical for everyone to be positioned close to a wall, so to allow a 5 m (16 ft) patch cord to be used otherwise the room would have everyone clustered around the edge with an empty space in the middle – like a giant donut.

Figure 10.4
Modular jack
wiring

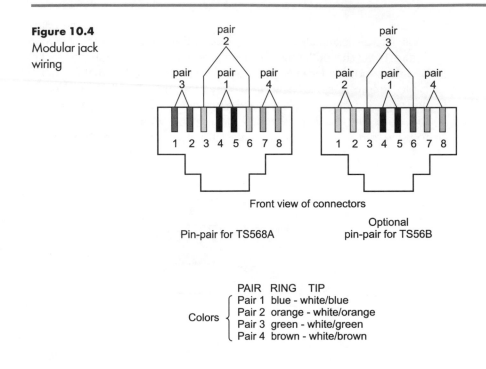

Front view of connectors

Pin-pair for TS568A

Optional
pin-pair for TS56B

PAIR RING TIP
Colors
Pair 1 blue - white/blue
Pair 2 orange - white/orange
Pair 3 green - white/green
Pair 4 brown - white/brown

The solution is to use the pillars as well as the walls or other permanent structures to provide groups of outlets to serve the surrounding areas. These groups of outlets are called multi-user telecommunications outlet assemblies (MUTOAs) and provide up to 12 telecom outlets (TOs). These multi-user outlets should be easily accessible and not in ceiling spaces, under floors or in pieces of movable furniture.

Even with these multi-user outlets, it may be a problem to restrict the length of patch cords to the 5 m (16 ft) limits used in separate offices and it is allowed to extend them, providing another part of the overall system is decreased in length to compensate.

One method of making connections in these offices without having the floor area littered with cables is to use a flat cable designed to pass under the carpets.

These cables, referred to as 'undercarpet telecom cable' (UTC), allow outlets to appear anywhere on the floor without worries about pillars or walls. It also allows last-minute flexibility in the office layout. They should only be used where other systems are impractical, as they generally have higher losses and the under carpet run should not exceed 10 m (33 ft).

Transition point

When the flat cable emerges from under the carpet it enters a connector box called a transition point (TP) where it is connected to round cable which involves less losses.

Undercarpet telecom cable (UTC) installation

As the cable is going to spend its whole life being pressed onto the floor, we must provide a smooth surface. If there are any holes, rough surfaces or grit on the surface it will tend to be pushed into the cable. Some protective tape is recommended under the flat cable which can be metallized if necessary.

It is often the case that we need power cables to follow a similar route and for undercarpet use, flat power cables are available. However, we should avoid running them parallel unless separated by more than 150 mm (6 inches) and avoid crossing the cables. If this is essential, it is better for the power cable to pass under the flat telecom cable.

Avoid wet conditions and where the cable may come into contact with solvents.

There are some disadvantages

It is not easy to make changes in direction to pass around obstructions and it is a good idea to consult with the cable manufacturer about their recommended method of doing this.

As the carpets become compressed by the foot traffic, the route taken by the undercarpet cables becomes visible and the cables may be damaged by wheels on heavy equipment being moved around.

Any modification of the cable route will involve lifting the carpet which, in itself, may need furniture to be moved. Carpet tiles can be used to reduce this problem.

Consolidation points (CPs)

In large office spaces it may be an advantage to have a connection point to link the 'horizontal' cabling involved with that area with the incoming 'horizontal' cable that we will look at in a moment. The overall layout of part of one of these large office spaces is shown in Figure 10.5.

We do not have to use consolidation points but if we decide to use them we need them in a position where they are easily and fully accessible. Something that is difficult to reach does not aid good workmanship. It can be inside floor or ceiling spaces providing that the location is

Figure 10.5
A mixed office
environment

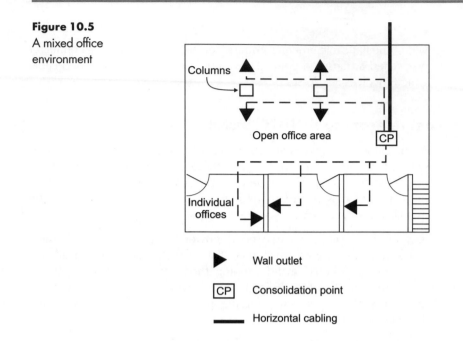

Figure 10.6
The route so far

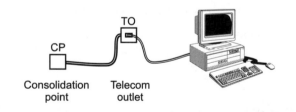

identified and it is fully accessible. We should not install any active telecom equipment in these spaces.

Consolidation points should not be used as a transition point.

The progress of our data from our computer is shown in Figure 10.6.

To the telecommunication room

From the work area, the cables pass though the telecom outlets and may pass through the optional transition or consolidation point and from here the cable enters the telecommunication room (TR) and terminates at a group of connectors which may be punch down blocks or a patch panel. These connectors are referred to as a horizontal cross-connect

(HC) – also called a floor distributor (FD). The telecom room is an enclosed space to contain the connections and other pieces of telecom equipment.

Topology

The recommended topology for horizontal cabling is a star topology, although other topologies like bus or ring can be used if required.

Cable used – types and lengths

Horizontal cables are usually four-pair 100 Ω 24 AWG 0.51 mm (0.020 inches) solid conductor unshielded pair cables or two-pair 150 Ω shielded twisted pair; Cat 5e or higher is recommended. They may also be two-fiber optic-fiber cables – just mentioned in case we meet one.

From horizontal cross-connect (floor distributor) to intermediate cross-connect (building distributor) must not exceed 300 m (984 ft).

From intermediate cross-connect (building distributor) to main cross-connect (campus distributor) must not exceed 500 m (1640 ft) unless the horizontal cross-connect (floor distributor) to intermediate cross-connect (building distributor) distance is reduced, in which case a maximum of 800 m (2624 ft) applies.

The above figures are from ANSI/TIA/EIA-568-B.1. In due course, the ISO/IEC 11801 will be available but are likely to be very similar.

Bridged taps

These are multiple appearances of the same cable pair at several distribution points and are not allowed in horizontal cabling. We like to find one end of the pair and then the other end so we can be certain that it is a simple link between the two sites. If we suddenly find another connection to the same pair – or several connections – we would have no idea what route the signals have taken. It would make faultfinding and circuit tracing unnecessarily difficult.

Give them some slack

At the cable end in the telecom room, allow for about 3 m (10 ft) of spare cable to allow for any slight repositioning in the future. Store this slack in a figure of eight pattern to prevent stress on the cable. If it must be coiled, keep it very loose and never bundled in a tight loop as this will degrade the performance of the cable. This also applies to any optic fibers that are found in the telecom room. At the other end of the cable where it enters to telecom outlet, allow a shorter length to allow for easy inspection and slight movements. A length of 300 mm (12 inches) will be sufficient.

The cabling at this stage will be similar to that shown in Figure 10.7.

Figure 10.7
Horizontal
cabling to two
work areas on a
single floor

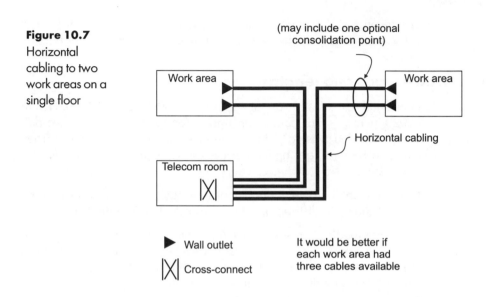

(may include one optional
consolidation point)

Work area

Work area

Horizontal cabling

Telecom room

|X|

▶ Wall outlet

|X| Cross-connect

It would be better if
each work area had
three cables available

Backbone cabling

The data from our computer has now have passed through the horizontal cable and entered the telecom room or the equipment room, as in Figure 10.8.

Backbone cabling is the connections between the telecom room and where the cable first comes into the building and between interconnected buildings. The topology used is either physical star or physical hierarchical star but these can be connected to provide other logical topologies as required.

Figure 10.8
Now we have
added the
horizontal
cabling

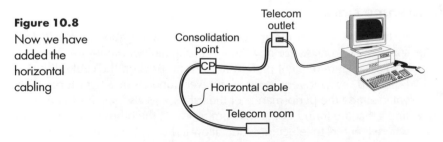

Telecom
outlet

Consolidation
point

CP

Horizontal cable

Telecom room

Bending cables

There are two points here: we must not bend and straighten copper cables too often and we must not install it with a very sharp bend.

With regard to the first problem, copper is work-hardening, which means that it becomes brittle if repeatedly bent and will break. Stranded copper is much more tolerant of repeated bending provided that it is not too severe a bend, which is why we always use stranded copper for equipment cables.

The minimum bend radius is always supplied by the manufacturers but generally we should not bend backbone cable so that the radius of the curvature is less than 10 times the overall diameter of the cable. For other cables, such as four-pair cable, the limit is four times the cable diameter.

Telecommunication room

This is an enclosed room or area where telecom equipment can be housed together with the connections between the horizontal cabling and the backbone. There may be several telecommunication rooms in a single building as it may be convenient to split the area into different zones.

Equipment room

This is another room that is used to house telecom equipment and other complex electronic equipment that is used to serve the whole building. It probably won't house the interconnections between the backbone and horizontal cables but it could do, so it is somewhere to look if we have lost the cable connections.

Cabling buildings

We will have a look at a campus situation where several buildings are interconnected, perhaps on an industrial site or within a university, and then we will look at a single multistory building.

Interconnected buildings

The telecom service enters one building and is connected to the backbone cable and fed into an equipment room where it joins the main cross-connect (MC), which is also called a campus distributor (CD) and is the center of the backbone star. From here the backbone is star-wired to connect with other buildings, as in Figure 10.9.

123

Figure 10.9
Differing
terminology
used in
interbuilding
cabling

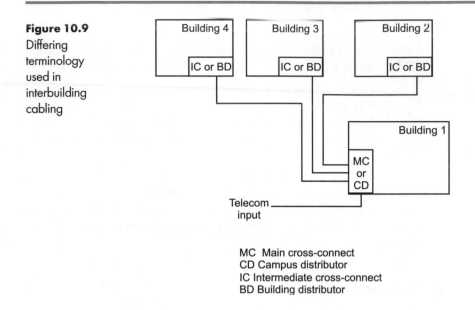

MC Main cross-connect
CD Campus distributor
IC Intermediate cross-connect
BD Building distributor

Inside each building it enters at an intermediate cross-connect (IC) also known as the building distributor (BD) or, in small buildings, directly to the horizontal cross-connect (HC) or floor distributor (FD).

We are getting a lot of names popping up – all these cross-connects, etc. It would be worth a few moments to get them sorted. In Figure 10.10 we can see the hierarchical structure and the alternative names. The US preferred names, all the 'connects', are on the left of the diagram, and all the international equivalents are on the right.

Multistory buildings

As we are tending towards a greater exchange of standards and because many companies now contract on an international basis, it often happens that we read information that has been prepared in another country, or even continent. One small point that may be worth mentioning at this point is that the naming of floors in a multistory building is far from consistent. In some countries, the floor which is at ground level is called the first floor, and the one above it is the second floor whereas in other parts of the world, the floor at ground level is referred to as the ground floor and the one above it is the first floor.

In a multistory building, the usual method is to bring the cable in at ground level at a main cross-connect (campus distributor) or intermediate cross-connect (building distributor) as the hub, then take the backbone to each floor to a horizontal cross-connect (floor distributor), as we can see in Figure 10.11.

Figure 10.10
Sorting out the
names

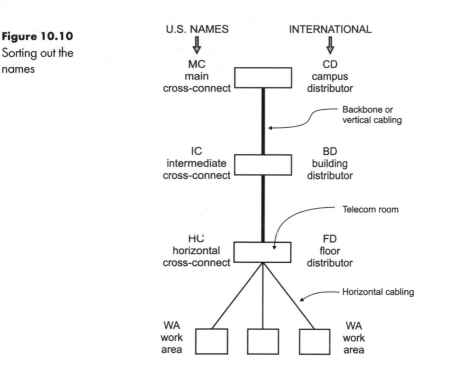

U.S. NAMES   INTERNATIONAL

MC
main
cross-connect

CD
campus
distributor

Backbone or
vertical cabling

IC
intermediate
cross-connect

BD
building
distributor

Telecom room

HC
horizontal
cross-connect

FD
floor
distributor

Horizontal cabling

WA
work
area

WA
work
area

Figure 10.11
Star topology
for a multistory
building

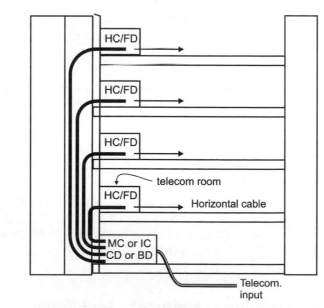

HC/FD

HC/FD

HC/FD

telecom room

HC/FD

Horizontal cable

MC or IC
CD or BD

Telecom.
input

MC Main cross-connect; CD Campus distributor;
IC Intermediate cross-connect; BD Building distributor;
HC Horizontal cross-connect; FD Floor distributor.

The connection to each floor is made in the telecom room and, wherever possible, these should be aligned vertically so the backbone cable can then be dropped down the side of a multistory building. This practice also explains why backbone cables are also called 'vertical cables'.

An alternative for a very large building that is using a hierarchical star is shown in Figure 10.12. The intermediate cross-connect (building distributor) and the horizontal cross-connect (floor distributor) can both be installed in a single telecom room on each floor.

Figure 10.12
Hierarchical star topology for a large multistory building

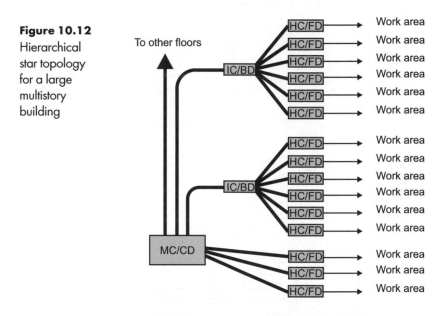

MC Main cross-connect; CD Campus distributor;
IC Intermediate cross-connect; BD Building distributor;
HC Horizontal cross-connect; FD Floor distributor.

Cable for the backbone

As expected, the choice of cable is between optic fiber and copper. The copper is 24 AWG 0.51 mm (0.02 inches), round, solid core conductors with a characteristic impedance of 100 Ω twisted pair. Existing backbones may also include 50 Ω coax cable or 150 Ω shielded twisted-pair (STP) cable.

Backbone cabling in multistory buildings

With the telecom rooms being placed one above the other on each floor, it makes it convenient to install a vertical cable through some sort of gap in the floor and each ceiling. These gaps go by three names, slots, cores and sleeves, though they are all basically just 'holes in the floor'.

An important safety point arises as a consequence of using this design. This system of holes in each floor, one above the other will give rise to a fire hazard as we have really built a chimney to carry smoke, fumes and fire from one floor to the next. We discuss the solution to this problem in Chapter 12.

These holes should be positioned close to the wall so that the cable weight can be supported by attachments to the wall but the holes should not be in a position that would obstruct the termination of cables within the telecom room.

Slots, cores and sleeves

Slots are rectangular holes through each floor, the size of which is determined by the area of the floor being served and indirectly by the cable to be housed. As a minimum, the dimensions are 150 mm (6 inches) × 225 mm (9 inches) for floor areas up to 25 000 m^2 (250 000 ft^2). The size increases progressively until floor areas of 140 000 m^2 (1400 000 ft^2) to 200 000 m^2 (2000 000 ft^2) suggest slot sizes of 375 mm (15 inches) × 600 mm (24 inches).

A circular hole cut through a floor for access is sometimes referred to as a core.

Sleeves are virtually the same except that they are round like a piece of conduit and usually of 100 mm (4 inches) diameter. Increasing floor area is accommodated by adding more sleeves rather than just by increasing the diameter of the sleeve. Small floor areas of up to 5000 m^2 (50 000 ft^2) employ three sleeves and large areas between 30 000 m^2 (300 000 ft^2) and 50 000 m^2 (500 000 ft^2) use between 9 and 12 sleeves.

Slots and sleeves should be designed into the building and, if they need to be cut after or during the construction stage, it is imperative that the structural engineer and appropriate authorities are consulted. We really don't want an ill-designed system introducing weaknesses, one above the other, in a multistory building.

Figure 10.13 shows typical slots and sleeves.

Open shafts

These can be used where available except if used as an elevator. Installations in elevator shafts are strictly forbidden on safety grounds both for access and because of the danger of fire.

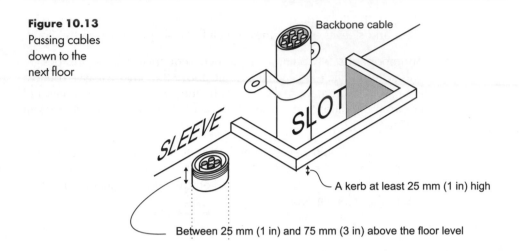

Figure 10.13
Passing cables
down to the
next floor

Backbone cable

SLEEVE

SLOT

A kerb at least 25 mm (1 in) high

Between 25 mm (1 in) and 75 mm (3 in) above the floor level

Installing heavy cables

We look at this in our safety section in Chapter 12.

Connecting two buildings

Between two buildings, the inter-building backbone is usually taken underground or by an aerial cable.

Aerial pathways

Aerial pathways can be installed quickly and easily being supported by poles, towers and the sides of the buildings. It provides easy access for maintenance but there are drawbacks. The easy access may be a disappointment if the cables cross property not owned by the organization.

There are disadvantages for the property owner in that it could damage the building and even if it doesn't, not many people think that trailing wires adds much to the overall appearance of the site. The cables are exposed to lightning, falling trees and other extremes of weather. We may also have to be careful about height clearance for traffic.

Underground cabling

Using pathways

These are a system of underground conduits with maintenance holes/ manholes and/or pull boxes that are normally installed by the building owner to feed cables between the various building entrance facilities.

Advantages of underground pathways:

(a) It looks nicer without cables and poles everywhere.
(b) It is economical over the lifetime of the system.
(c) The system can be modified without (too much) disruption to the buildings.

The disadvantages:

(a) It is slow and expensive to install.
(b) Careful planning is involved to prevent damage to utilities already installed.
(c) Unless carefully sealed, gas and water may be able to get into the buildings.

Using direct-buried cables

Using this method, as the name suggests, we bury the cable straight into the ground with any conduit.

To prepare the ground, we can use a trench digger to excavate a trench or vibratory plow that is able to cut a slit in the ground, feed the cable into it and then refill the slit all in one operation. When using any of these machines, we must take all steps to ensure that the ground is safe. We don't want to cut through any electricity, water or gas pipes – especially not all at once.

Another alternative is directional boring. This is a drill that can be controlled to follow a path or even go around an obstruction. This is very useful if trenching is not possible, like going underneath a building or a river.

Advantages of direct-buried cables:

(a) It looks better than aerial cables.
(b) It's cheap – at least initially.
(c) It can go around obstructions.

The disadvantages:

(a) The ability to go around obstructions means that no accurate path can be documented.
(b) It does not provide additional protection for the cable sheath.
(c) Not easy to modify the system at a later date.

Locating previously buried cables

Accidental method

We don't want anyone to find our cables by this method. It involves using a trencher or digger and accidentally hitting our cable. To reduce the chances of this, we should add a layer of warning tape at least

450 mm (18 inches) above the cable. Hopefully, this may alert the operator of the danger before damage results.

Electronic methods

We have two methods, the first is passive in that the cable does not cooperate in any way. We use a metal detector to detect the copper or the armoring of the cable. We may locate the warning tape if it happens to be metallic – it isn't always. The second method is able to detect cables at a greater depth. We apply a signal onto the cable which can then be detected from the surface.

Dig a hole method

We can use a high-pressure water or air jet to break up the soil above the cables, which can then be sucked out to leave a hole about 150 mm (6 inches) diameter. This process can continue down several feet until the cables are located. The water or air has no detrimental effect on any cables located.

A final thought

Installing cables is very expensive.

The cable is relatively cheap.

Everything takes more cable than we think.

So, round-up the figures, overestimate the length of cable required, then round-up the figures.

Chapter 10 quiz

1 Consolidation points connect:

(a) vertical cable to backbone cable.
(b) backbone cable to horizontal cable.
(c) horizontal cable to equipment cord.
(d) horizontal cable to horizontal cable.

2 An intermediate cross-connector is the same as a:

(a) horizontal cross-connect.
(b) building distributor.
(c) main cross-connect.
(d) floor distributor.

3   The topology most often used by a group of interconnected buildings is:

(a) ring.
(b) bus.
(c) star.
(d) double ring.

4   Elevator shafts:

(a) must NOT be used to contain telecom cables.
(b) can only be used to install vertical cable.
(c) must be positioned adjacent to telecom rooms.
(d) require a fire stop at each floor.

5   To reduce the chance of underground cables being accidentally dug-up it is recommended that we use:

(a) warning tape.
(b) yellow dye added to the soil above the cable.
(c) a warning notice fixed to the nearest permanent structure.
(d) a smoke canister with a trigger positioned at least 450 mm (18 inches) above the cable.

11

Does it work?

We need to test wiring to prove that we have done the work correctly and to locate any problems with the system. The first and most obvious check is to see if the cables go to the right places.

There is a series of tests to be carried out on cables. As we will see in a moment, some are applicable to all cables and some are applied to just a few, depending on category of use and whether they are twisted pairs or coaxial cables.

To help us, we have available a range of instruments. It is a good idea to read through a few cable suppliers' catalogs to see what is currently available.

Acceptance test is the name given to the final test of a new or upgraded system that normally takes place after a short period of use, called a burn-in period. This is the final check to show the customer that the system is working as ordered. Minor errors can be accepted and sorted out later but serious ones will have to be corrected before acceptance.

Not all tests mentioned in this chapter form part of the acceptance test for a particular cable as they may be more applicable to faultfinding procedures but we will itemize the essential tests in a minute.

Test equipment

Instruments vary in price and sophistication. Most testers work by connecting a unit to each end of the cable and sending tone signals along the cable, making measurements at either end according to the tests being made.

We can buy very cheap tone generators to do this job or we can pay real money for a fully automatic system that will perform all the tests, including wire mapping and the others to be mentioned in this chapter, and within seconds can provide a pass/fail indication for every required test. They will even store a few hundred test results, so all the checks made on a full system can be downloaded at the end of the day to keep in our records and to give a copy to the customer. Top of the range test equipment is always available for lease so that we do not have to invest too much money into equipment that we may not use very often before it is overtaken by new models.

We pay for accuracy and this is particularly true when we are using high-frequency systems, so we must choose instruments that comply with the current standards.

There are three levels of accuracy required by instruments:

- For Category 5 cabling, level II instruments are required for field acceptance testing.
- For Category 5e cabling, a level IIe instrument is required.
- For Category 6 cabling, a level III instrument will be needed.

Wire map testers

These instruments have a main unit that is connected to each of the wires in a twisted-pair cable. It sends signals through each wire to another unit connected to the far end of the cable. In this way, it can immediately check to see that the wires have been connected correctly. This is called wire mapping and will be considered in a moment. Their simple function is reflected in the low cost of the tool. This facility is also built into more expensive and sophisticated pieces of gear.

Tone generator and detector

This is a simple continuity tester. We go to one end of a wire and clip on the tone generator, which injects an audio signal between about 500 hertz (Hz) and 5 kHz onto the wire. The detector unit is like a small hand-held probe which can be touched onto the far end of the wire and a built-in speaker tells us that the cable has no breaks in it. Since it is small, light and hand-held, we can easily follow the wire around the building checking that the circuit is complete. By touching it onto adjacent wires, we can check for short circuits. It's not rocket science but it's quick and easy.

Cable analyzers

Unlike the tone generator, these really are rocket science. They are hand-held and offer a very wide range of capabilities. They perform complete acceptance tests on a wide range of cables in 10 s, monitors traffic on the lines up to 350 MHz, shows detailed plots of NEXT (near-end crosstalk), ELFEXT (equal level far-end crosstalk), ACR (attenuation to crosstalk ratio) and other characteristics. They also provide a two-way voice communication between the main and remote units. This is very useful and saves a lot of shouting and running. Finally, it provides a Windows® compatible output so we can download and print off copies of all our results.

Many of these have a time domain reflectometer included in the (substantial) price.

Time domain reflectometer (TDR)

This instrument can test any cable, whether the conductors are copper, aluminum or steel – in fact any conductor. It is not specified as being essential for acceptance testing for twisted-pair cable but is used for coaxial testing, faultfinding and general testing. It used to be a bulky piece of kit but, as with everything else, it has now shrunk to hand-held proportions and is getting cheaper. It is a very clever device.

There is a similar device called an optical time domain reflectometer (OTDR) that performs the same functions on optic fibers.

Figure 11.1 gives an idea of what is inside. It all starts with the timer. When the timer shouts 'go', a very short duration voltage pulse is put onto the cable under test. At exactly the same moment a trace starts moving across the display. The TDR is adjusted so that the trace moves across the screen in the same time as is taken for the transmitted pulse to get to the end of the cable, and back.

Figure 11.1

The brains of a TDR

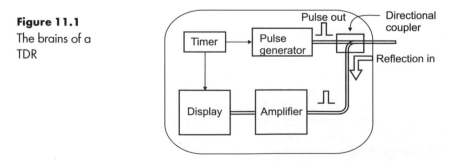

The pulse travels along the cable and every time the impedance of the cable changes some of the energy is reflected back towards the TDR. A reflected pulse enters the TDR and is redirected into an amplifier and then to the display. Each reflected pulse is shown at a point on the screen corresponding to its position along the cable, so we get a sort of picture or map of the cable showing any interesting events.

In Chapter 3 we saw that a change in the characteristic impedance of any cable will cause some reflected energy to be returned back along the line. If we had a continuous line with the end terminated perfectly, then all the energy would be absorbed and we would have no reflections. The bigger the change in impedance, the larger the reflected energy, so even tiny changes like a connection between two cables would give a small amount of reflected energy The TDR on the end of a cable can 'see' every change of impedance along the length of the cable.

How do we know the difference between an open circuit and a short circuit at the end of a cable? Assuming we apply a positive-going pulse to the cable, a short circuit reduces the voltage (theoretically to zero). This fall, or negative-going change, would be reflected back to the TDR whereas an increase in impedance, like an open circuit, would cause an increase or positive-going change. So, at the TDR, we see negative pulses for each reduction in impedance and positive pulses for increases in impedance.

As the reflections from more distant points will arrive later, we can display all the changes on a screen as a graph with distance along the bottom and amplitude of reflections displayed vertically, as in Figure 11.2. There will always be some small signals being returned due to electrical noise in the TDR, slight amounts of electromagnetic interference (EMI) and slight changes in cable characteristics so the horizontal line is never absolutely straight.

Figure 11.2

A cable under test

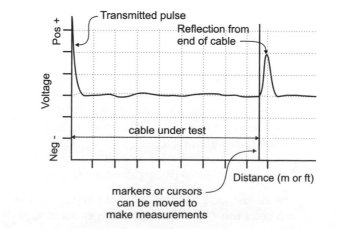

In the figure, we see a marker or cursor that can be moved across the display to take measurements of any point of interest. We have a reflection from the start of the cable and then another showing the reflected energy from the end of the cable. Notice how measurements are taken from the beginning of the spike as this is the actual point where the reflected energy starts appearing. The spike height just shows us how much energy is being reflected or by how much the impedance has changed.

Notice how the polarity of the transmitted pulse and the reflection are the same. This is a sure sign that the impedance has increased. In Figure 11.3 we see the same cable with the far end short circuited. Whenever the impedance is reduced, the polarity of transmitted pulse and the reflection are always opposite.

Figure 11.3
A short-
circuited cable

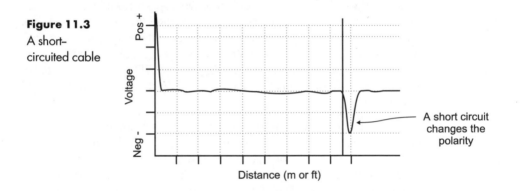

All TDRs offer the facility of different lengths of transmitted pulse. The values may vary from very narrow pulses of 2 ns up to about 2 μs. Why is this? Well, the clue is in the cable length that is to be measured. They cover a wide range, typically from 1 m to 3 km (3 ft to 10 000 ft), which is quite a range.

Now, if we are to get a reflection from the end of a cable, we need to send a signal strong enough to be reflected off the far end and travel all the way back along the cable and still be large enough to be displayed on the screen. In everyday talk, we often refer to this as the 'power' of the TDR, so we need a more 'powerful' one to check long cables. However, what we are really referring to is high energy contained in the pulse which is proportional to the length or width of the pulse. The voltage is usually fixed, let's say 5 volts (V), and the time is the length or width of the pulse.

The longer pulse will contain more energy so, like shouting louder, we can get a useable echo from a longer distance.

So why not just use the longest pulse all the time? If the voltage pulse travels along the cable with a nominal velocity of propagation (NVP) of 0.7, it will be moving at (very roughly) 0.2 m or 8 inches in a nano-second, so a 10 ns pulse would extend for 2 m (6.6 ft) along the cable. If two things or 'events' occur on a line, say, 1 m (3.3 ft) apart, a 10 ns pulse would manage to hit both at the same time and the reflections would merge into one and so the TDR would not be able to show the two events separately. To see them as two events we would have to use a shorter pulse length. By reducing the pulse length to, say, 0.2 ns the two events would be displayed separately.

Pulse length is therefore a compromise between long pulse for maximum range and short pulses for the separation of details so, for short ranges we use a short pulse and as the range increases so must the pulse length. Figure 11.4 shows how events can be lost when using longer pulse lengths. The flat tops to the waveforms for medium and long pulses are caused by the amplifier being overloaded by the strong reflections.

Figure 11.4
The effects of changing the pulse length

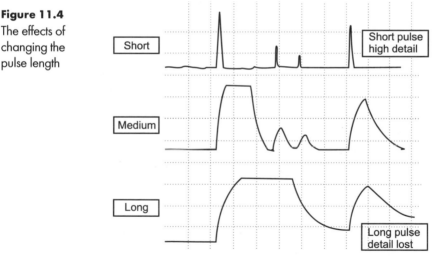

How do we know what things are?

In the previous figure, we have two pulses showing on the TDR screen. We cannot glance at the screen and immediately say what they are but all is not lost – we do know some things about them. We can tell the distance from the start of the cable and we can tell their distance apart. What else? The pulse has the same polarity as the launch pulse so we know that these two things that have caused a sudden increase in the impedance of the cable. The cable may be partially open circuit – perhaps it has been damaged during installation. Perhaps the cable has been reconnected at those points, possibly untwisted too much when being terminated. Finally, if we have an over-suspicious disposition we could think it could be some bugging device secretly tapping into our communications, which, I suppose, is slightly more likely than aliens beaming up our data.

The only thing to do is to investigate and when the cause is discovered, make a copy of the TDR trace (most TDRs give us a printout or download facility) and note the cause so that we can build a bank of experience to help us next time.

Finally, in Figure 11.5, a trace showing the effect of radio frequency signals being present and water penetration into the cable. Notice that the water causes a downward slope at the beginning and an upward slope at the end of the water.

Figure 11.5
Two other TDR traces

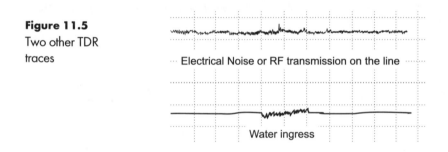

Acceptance tests

Field testing of horizontal twisted-pair cables is based around standard layouts devised by the TIA/EIA and the IOC/IEC 11801.

The basic link

This configuration by TIA/EIA TSB67 is shown in Figure 11.6. Notice that this includes the test leads to the test equipment.

Figure 11.6
The 'basic link' test configuration

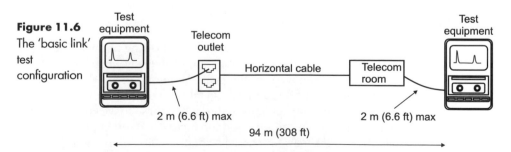

The permanent link

This is the ISO/IEC 11801 (almost) equivalent to the basic link except for the lack of test leads, which result in slightly different criteria for the test. This is shown in Figure 11.7.

Figure 11.7
The 'permanent link' test configuration

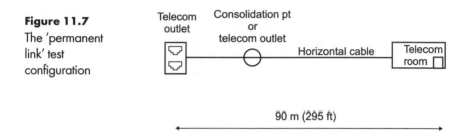

The channel

This layout is common to both authorities. It is an overall, bit of every-thing, link which, as we would expect, is a little longer – see Figure 11.8.

Figure 11.8
The 'channel' test configuration

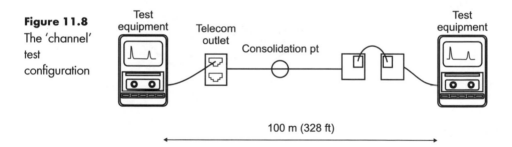

The current tests for all categories of twisted-pair cable are:

1. Wire mapping.
2. Length.
3. Attenuation.
4. NEXT.

In addition, Cat 5 (and Class D) and higher require:

1. ELFEXT.
2. Delay and delay skew.
3. Return loss.

Also, Cat 5e and higher require PSNEXT and PSELFEXT.

Testing the cables

Wire mapping

This is really just a matter of checking to see whether all the wires are connected properly and that nothing disastrous has happened, like a disconnection. This is a required acceptance test.

With a coaxial cable, we would have to work quite hard to get it wrong as there are only two conductors – the thick braid that goes around the outside and the thin center connector. Since these are very different, it would take only a moment to check how they are connected. The more conductors that we are dealing with, the more likely we are to have an error.

If we take twisted-pair cable as an example, we can see in Figure 11.9 the correct wiring pattern that we would expect. Remember that T568-B and T568-A use the same colors but pairs 2 and 3 are reversed – as they are both connected straight through it will still work OK but it is better to keep everything in an installation consistent with one standard.

Figure 11.9
Everything as it
should be

Cable faults

These faults are either manufacturing faults – highly unlikely, or damage inflicted on the cable during installation – more likely. It may be caused by pulling too hard or bending the cable too sharply or repeatedly or dropping something heavy on it. Remember that copper 'work hardens' with bending. This has the effect of making the copper more and more brittle as it is repeatedly flexed. These forms of damage may not be obvious just by looking at the cable.

The first possibility is a break in one or more of the conductors. A disconnection or break is called an 'open' circuit. An open circuit may also occur in the shield continuity in shielded cables. This will reduce the effectiveness of the shielding.

The alternative is that the insulation fails and two or more cores can touch and we have a 'short' circuit. It could be caused by bending but it is more likely to result from careless stripping or termination of the cable. We can have a short circuit across a single pair or from one pair to another or from a conductor to a ground point.

These faults are shown in Figure 11.10.

Figure 11.10
Opens and
shorts

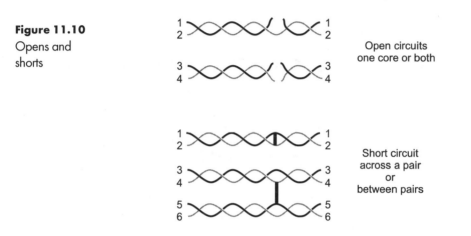

Open circuits
one core or both

Short circuit
across a pair
or
between pairs

Wiring faults

These are really down to the workmanship of the installer. With eight wires in a four-pair cable, there are lots of ways to get it wrong but they are all color coded so, with a bit of care, it should be easily avoided.

Reversed pairs are a simple matter of getting the two wires in a single pair reversed so the tip and rings are reversed. Transposed or crossed pairs are similar except in this case, we have swapped two pairs over, so pair 1, for example, are connected to the terminals that are expecting pair 2.

Split pairs are not easy to locate as simple tests at each end of the cable will not show up any problem. Indeed, if we follow the cores in the figure, they go to their correct terminals, so it would appear that the wire map does not show a fault. The correct connections appear at both ends of the cables and can only occur if the same fault has occurred at each end of the cable. The idea of twisting the wires is to reduce cross-talk by having two adjacent conductors carrying equal signals in opposite directions and thus canceling their electromagnetic fields.

When we have split pairs, the connections of one wire in each of two adjacent pairs is reversed at both ends so we have two pairs of wires twisted together that are carrying different signals and so their electromagnetic fields do not cancel.

Low frequency testing would not cause much crosstalk and would be unlikely to be noticed but as the frequency increases problems would increase, so if we have complaints about a communication link that checks out OK, then this is something to think about. Some automatic testers can spot this but others can't.

These faults are shown in Figure 11.11.

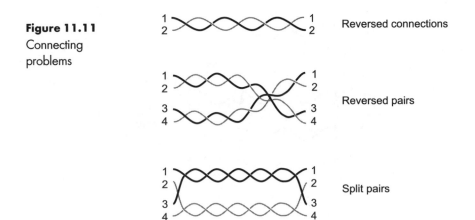

Figure 11.11
Connecting
problems

Reversed connections

Reversed pairs

Split pairs

Test requirement: no mistakes – it all has to be OK.

Direct current (DC) loop resistance

By short circuiting an end of a pair of conductors, a resistance meter or ohmmeter can be used to check the total, there and back, resistance of the cable as in Figure 11.12. By comparing the result of other pairs in the same cable, any wide differences should give cause for concern.

Test requirements: not required for twisted-pair cables but usually required for coax, depending on the application.

Figure 11.12
Measuring
'there and
back' resistance

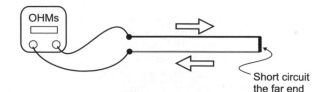

OHMs

Short circuit
the far end

Length of a cable

A cable has two lengths: physical length and electrical length. The physical length is easy, we just lay the cable out and measure the length of it or read off the length measurement off the jacket. The electrical length if the distance traveled by the signal from one end of the cable to the other, which at first sight would appear to be the same, but the conductors inside the cable are twisted. As we twist the conductors this uses up distance, so the electrical length is greater than the physical length.

To reduce crosstalk, the rate of twist of each pair is different and so the electrical lengths will work out differently within the same cable. The electrical length of the shortest pair is used as the cable length in the regulations, so some of the other pairs may have electrical lengths greater than the regulation maximum (Figure 11.13).

Figure 11.13
The tighter the twist, the longer the cable appears

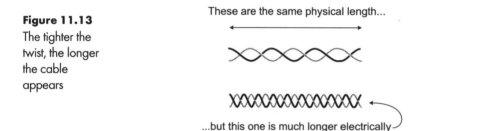

These are the same physical length...

...but this one is much longer electrically

So, how do we find the electrical length of a cable? We use a test instrument that sends a short pulse of voltage along the cable to the end which is left open circuit. The instrument then 'listens' for some energy bounced off the end of the cable and measures the time it takes between sending the pulse and getting the reflection back, as shown in Figure 11.14. So, all we need to know is how long the pulse took, and how fast it was traveling. Easy.

Figure 11.14
Finding the travel time

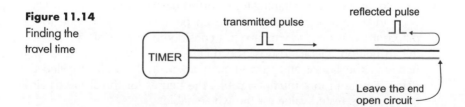

transmitted pulse

reflected pulse

TIMER

Leave the end open circuit

So how do we know how fast the voltage pulse can travel? The easy way it to ask the manufacturer – it's not a secret and they will be happy to tell us. Most instruments are loaded with data on each type of cable, so we just punch in the cable details and out pops a speed. The speed at which the voltage moves is often quoted as a fraction of the speed of light, which is nearly 300 m per microsecond (984 ft per microsecond).

Most instruments input the travel speed as a fraction of the speed of light rather than as an actual figure. This value is called the nominal velocity of propagation (NVP), so a speed of 0.7 means 0.7 of the speed of light.

$$\text{cable length} = \frac{\text{NVP} \times c \times \text{total travel time}}{2}$$

where c is the speed of light.

The total travel time measured is the time taken for a short pulse of voltage to travel to the end of the line and be reflected back. This also accounts for '2' that appears on the bottom line. We had a look at reflections in Chapter 3.

Why do we have to use a reflection? By using the reflected signal we don't need additional timing equipment at the far end of the cable with the nightmare problem of 'starting the clock' at the same moment that the signal leaves the sending end.

What if it is an unknown cable? We have to start with a known length of cable then punch in an NVP value and see what cable length we appear to have. If, for example, we know we have 100 m (324 ft) of cable and, after choosing an NVP of 0.7 our instrument calculates a length of 90 m (292 ft), we know that the speed is higher than 0.7, so we put in a higher figure and try again. Eventually we will obtain a calculated length of 100 m (324 ft) and so the last NVP guess must have been the right value for the cable. We can now use this NVP to measure the lengths of any cable of this type.

Test requirements: the lengths should not exceed those shown for the standard layouts in the figures except that an extra 10% is allowed to compensate for measuring errors, usually due to NVP errors.

Propagation delay and delay skew

Our test instruments can measure the time it takes to send a signal along a cable. In a cable containing four twisted pairs there is always a difference in the propagation time for each pair. This may be due to the different twists rates as we have just considered or possibly a change in the insulation characteristics. The difference in the propagation time between the fastest and slowest pairs taken at 10 MHz is called the delay skew and in a multipair cable, the figures for the slowest pair is taken as the tested value for the whole cable.

These tests are only applicable for Cat 5 and above.

Test requirements: using a 10 MHz signal on a 100 m (328 ft) cable, the minimum propagation speed is an NVP of 0.611 (61% of the speed of light) and the maximum delay skew is 45 ns.

Cable attenuation

This is the loss of power in the signal as it travels along a cable. It is made up of power loss due to resistance within the cable and radiated energy. Its value is usually quoted in decibels per 100 m but is highly affected by the frequency in use. For example, a cable may be quoted as having attenuations of 2 decibels (dB) for 100 m at 1.0 MHz rising to 41 dB at 300 MHz, so figures quoted without the frequency used for the test are quite meaningless. It is measured by using a calibrated transmitter at one end to send a known signal and a calibrated receiver at the other end of the cable to measure how much is remaining at the far end.

The input voltage and output voltage must be measured with exact matching at the far end of the cable. Given that this is difficult to achieve, we are really measuring insertion loss rather than exact attenuation of the cable. The formula is

$$\text{Decibels} = 20 \ \log\left(\frac{V_\text{i}}{V_\text{o}}\right) \text{dB}$$

The attenuation increases very quickly with frequency and slightly with temperature. The attenuation figures for a basic link configuration at 10 MHz are 10 dB for Cat 3 and 7 dB for Cat 5e.

Attenuation results for short links with a result of 3 dB or less produce unreliable results and should be disregarded.

Insertion loss

This is an easy one. It is the loss caused by anything being inserted in a system.

It could be the increase caused by inserting a plug and connector into a circuit or by inserting a whole cable, so it is more generally applied than cable attenuation, which is a measure of the characteristics of the cable alone.

Insertion loss includes all the losses involved with attenuation, connection and reflections, so the final losses are always greater than for just cable attenuation. Insertion loss is therefore always greater than attenuation. Test results for permanent link and channels are shown in Table 11.1.

145

Table 11.1 Channel and link data

Channel cable (category/class)	Insertion loss	NEXT	ELFEXT	Return loss	ACR
5e/D @ 100 MHz	24.0	30.1	17.4	10	6.1
6/E @ 100 MHz	21.7	39.9	23.3	12	18.2
7/F @ 100 MHz	20.8	62.9	44.4	12	42.1
Permanent link					
5e/D @ 100 MHz	20.4	32.3	18.6	12	11.9
6/E @ 100 MHz	18.5	41.8	24.2	14	23.3
7/F @ 100 MHz	17.1	65.0	46.0	14	48.7

NB. ACR is not specified by the Telecommunications Industry Association (TIA).

Return loss

As we saw when looking at the TDR, any change in impedance causes a reflection of energy. Return loss must be measured from each end of a cable and low loss links of less than −3 dB should be disregarded as the termination of the link will be having a significant effect. The return loss (R) is the ratio of the reflected voltage to the transmitted voltage and as the ratio is always less than one, when it is converted to decibels the result is always negative. The formulas and other technical bits are in Chapter 3.

Crosstalk (XT)

If a signal is passed along a conductor, the changing magnetic field can cut a nearby cable and induce voltages into it. The second cable then has a copy of the original signals superimposed on whatever signals are being carried. In some cases with analog signals, like a telephone conversation, it may well be audible in the other circuit. In other cases, the effect is to simply increase the noise level.

Remember that crosstalk is an electromagnetic effect and has nothing at all to do with electrons passing through the insulation as we sometimes see suggested. There are several forms of crosstalk measurement, depending on how the measurements are taken and these options are described in Chapter 8.

NEXT loss must be tested from both ends of the cabling. It is the cross-talk sum of the cable and near-end connectors – just one for the basic link and two for permanent links and channel configurations. As we would expect, the losses increase with frequency (see Table 11.1).

PSNEXT is a calculated value of the crosstalk occurring but is based on the measurements taken for the NEXT test (see Table 11.1).

ELFEXT loss is the crosstalk sum of the cable and near-end connectors – just two for the basic link, two or three for permanent links and four for channel configurations. Once again, it must be tested from both ends of the cable. The losses become greater as the frequency increases (see Table 11.1).

PSELFEXT is the crosstalk occurring from many different transmitters at the near end and measured at the far end (see Chapter 8). Once again, the losses become greater as the frequency increases (see Table 11.1).

ACR is the difference between the attenuation and crosstalk measurements for a twisted-pair cable. It is a calculation rather than a direct measurement. It is not required as an acceptance test for twisted-pair cabling. It is explained more fully in Chapter 8 and limits are given in Table 11.1.

Chapter 11 quiz

1   **A basic link has a measured loss of 10 dB. If the input signal had a value of 5 V, the output voltage would be approximately:**

   (a) 0.5 V.
   (b) 0.602 V.
   (c) 1.58 V.
   (d) 15.82 V.

2   **Insertion loss is:**

   (a) always greater than attenuation.
   (b) a calculated figure and cannot be directly measured.
   (c) always less than attenuation.
   (d) just another term for cable attenuation.

3   **A TDR used for a detailed examination of a short cable would use:**

   (a) the longest pulse available.
   (b) a very high voltage pulse.
   (c) the shortest pulse available.
   (d) only a negative-going pulse.

4   **Wire mapping is:**

   (a) the drawing up of the overall cable plan on a building.
   (b) only used with star topology.
   (c) the checking of cables for the correct connections.
   (d) a sketch of the core layout in a multipair cable.

5 A 'burn-in' period is:

(a) the period of fire resistance for sheathed cable.
(b) a short period of use after installation or repair.
(c) the time taken for a new set of regulations to be established.
(d) the expected lifetime of a cable installation.

12

Staying alive until payday

Think first

My colleague had just shown a film on industrial safety to a group of apprentices and was busy rewinding the 16 mm film back onto the empty spool. Now these spools really shift and when the film came off the spool, Jim put his hand on the empty spool to stop it. The metal spool sliced into his hand. Neither Jim nor the apprentices forgot the title of the safety film – 'Think First'.

We all suffer from such brain failure from time to time, avoiding an accident at work or in the street is more by luck than anything else. Sometimes we even know that it is going to happen and still carry on until it does. I remember scratching a line down a tile to help me to break it. The glazed surface was very difficult to penetrate with my knife so I was pushing hard. As I started, I remember thinking to myself 'if the knife slips you are going to have a nasty cut' but no sooner had I decided that it wouldn't happen to me then, of course, it did and I wasted the rest of the morning at the hospital.

Experience helps

We tend to learn by experience but it is a good idea to use other people's experience as much as possible. Sometimes, as I found, we know we are likely to have an accident and sometimes we don't. We may not realize that something is dangerous until it is too late.

In each country, and internationally, there are organizations and codes dedicated to keeping us safe, often by guarding us against ourselves. They seem at first glance to consist of endless regulations and restrictions and quite uninviting to read but, in reality, the regulations are the sum total of experience on committees that know more about the types and causes of accidents than we can (hopefully) experience in a lifetime.

In the States, for example, we have the National Electrical Safety Code (NESC) produced by the Institute of Electrical and Electronics Engineers (IEEE) and the US National Fire Protection Association (NFPA); in Canada the Canadian Standards Association (CSA) produces the Canadian Electrical Code (CEC); in Europe the European Committee for Electrotechnical Standardization (CENELEC) produces codes on grounding and bonding in buildings, and in the UK safety comes under the Health and Safety Executive (HSE). These are just a small fragment of the organizations and codes that deal with every aspect of workplace safety and other matters.

Our responsibility

We must find out what local regulations and codes apply to the job in hand and make sure they are complied with. This may extend beyond the codes applicable in the state or country in which the work is being carried out.

In most areas we are responsible for our professional practices and for our own safety and the safety of others. This can have far reaching effects. If we see someone breaking a regulation by taking a dangerous short cut or by smoking a cigarette in a forbidden area we can be held equally responsible if we don't take steps to prevent the safety hazard. Safety is not 'someone else's job' – like it or not, it is ours.

Unpleasant things that can happen to us

Electrical injuries

'It's volts that jolts but mils that kills.'

There is a lot of truth in this little rhyme. People have died from voltages of less than 50 volts (V) yet we all survive voltages of over 20 000 V on a regular basis with no lasting effect. We can easily pick up this type of voltage by walking on a carpet and then discharging it to give ourselves a nasty jolt when we touch something. This amount of charge can only sustain the current for a brief instant and does us no harm.

The 'mils' or milliamps (mA) is the thing that does the damage and the only connection with voltage is Ohm's law, $I = V/R$, so the real unknown is our resistance.

Anything that decreases our resistance and that of the circuit through which the current will flow can cause a problem – likely candidates are dampness or wet hands. Wearing jewelry is best avoided. A ring on the finger not only provides a good electrical contact to our finger but can firmly anchor us to a terminal or piece of machinery and putting on a necklace will also be connecting a good conductor around your neck. To some extent, the danger level is also affected by factors beyond our control such as skin thickness so it usually happens that children and women have lower tolerance to electric current.

Particularly dangerous are currents that flow through the chest area so we should avoid current paths that go from hand to hand or from hand to feet. If it is essential to work with the power connected, it is a good idea to work with one hand in a pocket to prevent the other hand from holding on to a good ground point (see Figure 12.1).

Figure 12.1
Currents to avoid

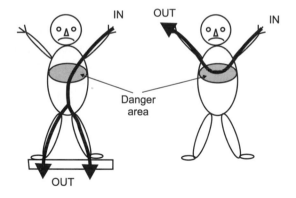

The safe limit is about 5 mA, at which point we will tend to leap back and throw ourselves off the contact point but this in itself may cause injury.

At higher levels of current between 10 and 15 mA our muscles contract and we may be unable to disconnect ourselves. Around 18 mA the diaphragm contracts and prevents breathing and we are liable to die of suffocation.

If the current increases over 30 mA our heart is likely to go into ventricular fibrillation. This is a very fast but uncoordinated series of contractions of the heart muscle that effectively stops the blood circulation (see Figure 12.2).

Figure 12.2
(Very)
approximate
current limits

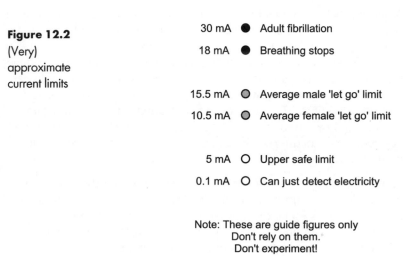

30 mA ● Adult fibrillation

18 mA ● Breathing stops

15.5 mA ○ Average male 'let go' limit

10.5 mA ○ Average female 'let go' limit

5 mA ○ Upper safe limit

0.1 mA ○ Can just detect electricity

Note: These are guide figures only
Don't rely on them.
Don't experiment!

With alternating current, the frequency is a significant feature. Low frequencies and high frequencies have less effect than the power frequencies between 50 and 60 hertz (Hz), as shown in Figure 12.3. At high frequencies, there is an increasing possibility of burns rather than electrocution. Burns are also the result of I^2R heating due to current flow just as we would get increased power dissipation in any other resistance.

Figure 12.3
Frequencies to
avoid

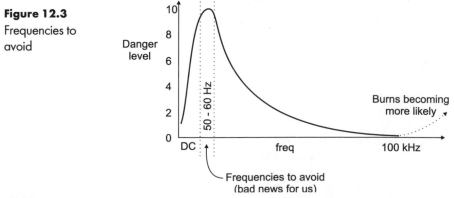

152

First aid for electrocution

The big problem with electricity is that the wire that can kill us looks exactly like a wire that is completely safe. The same applies to someone in contact with electricity. So ...

Do NOT touch the casualty with bare hands – you will be connecting yourself into the same electrical circuit and are likely to be electrocuted yourself and then there would be two casualties.

It is a sensible precaution for all of us to attend a training course on resuscitation, particularly those of us that use electricity as a living. It is nice to know that we are working with people who could possibly save our lives.

Electrical injuries are often internal and have no obvious symptoms even to the casualty, so we must get medical assistance however much the person insists that there is no problem.

I knew such a case. He was wiring a mains power socket at home and received a shock but became disconnected from the supply. He found himself laying on the floor and said to his wife, 'It got me but I'm OK'. He stood up and died of heart failure. It happens so easily.

Only suitably qualified and, and in some situations, licensed personnel should work on electrical circuits. Accidents in this area occur so quickly and easily with electricity that being 'fairly sure' is nowhere near enough.

Installation issues that affect us

Lightning

When we have a potential difference between two conductors separated by air, the electrons in the air molecules will be attracted by the positive terminal and repelled by the negative. This causes a slight shift in the orbits of the electrons but current does not flow until the potential reaches a critical value that will cause the electrons to break their bonds and a spark will occur. This happens in air at about 30 kV per centimeter (76 kV per inch) separation (see Figure 12.4).

If we have a build-up of electrical charge in a cloud, the potential difference will increase until the voltage will tear electrons out of orbit and a spark will jump between the cloud and the Earth's surface.

One thing we can try is dissipating the charge before it builds up to a dangerous level. A metallic spike can be used to discharge the area immediately above the spike by providing a resistive path between the cloud and the ground and, to some extent, dissipate enough charge

Figure 12.4
Electrostatic
breakdown

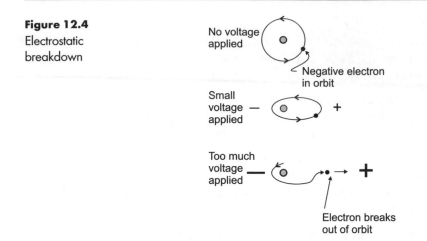

No voltage
applied

Negative electron
in orbit

Small
voltage —
applied

+

Too much
voltage —
applied

+

Electron breaks
out of orbit

to reduce the likelihood of a strike. It is not to be relied upon but if it helps, that is a good thing.

The path taken by the spark will be determined by the build-up of the electrostatic lines of force. These lines will be increased by a reduction in the distance between the cloud and the ground by buildings, people or trees, as shown in Figure 12.5.

Figure 12.5
The electric field
in a
thunderstorm

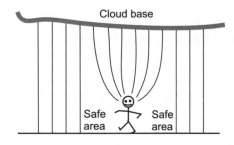

Cloud base

Safe
area

Safe
area

An object attracts electric lines of
force and possibly a lightning strike

The electric field can be diverted by offering it an easy route to ground by adding air terminals or lightning conductors. By diverting the electric field we create an area around the structure that do not contain any (or many) lines of electric flux and hence are extremely unlikely to be struck by lightning. This area is roughly circular and is called the 'cone of protection'. As a general rule, and far from a 'golden rule' the cone of protection has a radius of about the same as the height of the air terminals or lightning conductors. It's a working figure but lightning always springs surprises and behaves differently in different locations so we should always obtain local statistics wherever possible. The cones of protection for different-sized buildings is shown in Figure 12.6.

Figure 12.6
Safe areas around a building

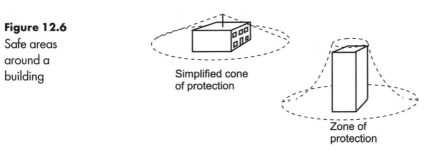

Simplified cone of protection

Zone of protection

Protection for a building can be achieved by enclosing the building in such a way that a steel structure surrounds the accommodation area – a bit like a Faraday's cage or the braid in a coax cable. To do this, we fix air terminals (lightning conductors) around the outside of the roof area and down the outside and into the ground (see Figure 12.7).

Figure 12.7
Conductive framework for a building

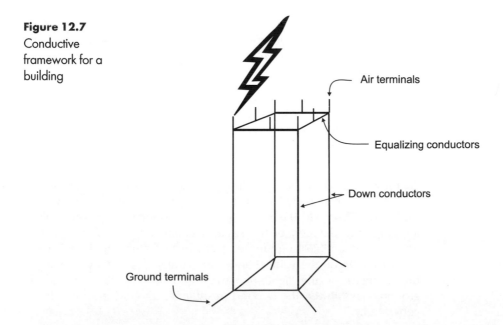

Air terminals

Equalizing conductors

Down conductors

Ground terminals

The lightning strike

Once the lightning has been intercepted, the current increases rapidly at an average rate of 10 000 A/μs – that's 10 billion amps per second! Some going.

It reaches a maximum value of between a few hundred and a few tens of thousands of amps with a statistical average of 20 000 A and there-after it decays at a much slower rate, often around a tenth of the speed, so a lightning strike follows the pattern shown in Figure 12.8. This strike would be referred to as a 1/10 strike, showing the ratio of the increase/ decrease times.

Figure 12.8

It's all over so quickly

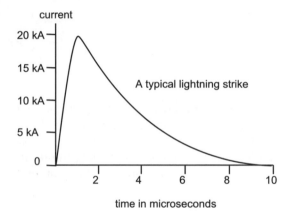

It may involve just a single strike, which would be all over in less than a millisecond, or a series of strikes that follow the same path over a period of a second or so.

As the point of strike on the lightning protective system may be raised to a high potential with respect to the true ground, there is a risk of a side flash, that is, a flashover between the air terminals and down conductors and metal on or inside the building. And the problems do not end there.

As these high currents enter the ground they will raise the ground potential to increasing values as we approach the grounding elec-trode. As Figure 12.9 shows, this can produce dangerous voltage levels for the duration of the strike. Animals and cows in particular, are very susceptible to injury or death in this situation. There are no figures shown on the vertical and horizontal scales as they depend on the strength of the lightning strike and the conductivity of the ground. The lower the ground resistance, the lower the voltages generated.

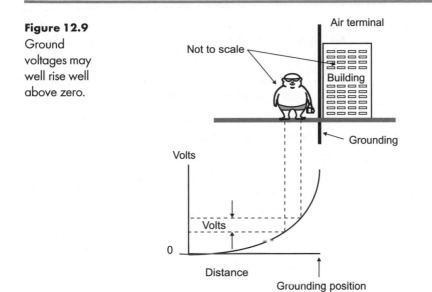

Figure 12.9
Ground voltages may well rise well above zero.

Finally, remember to use local statistics and local experience – lightning varies considerably in different locations. Your local authorities will have statistical analysis of the number and severity of strikes in your locality.

Underground cables

Burying the cables does not offer automatic protection against all ground strikes. The results depend on the resistance of the ground but a strike occurring within 2–6 m (7–20 ft) will still impact the cable. However, cables running between large buildings are generally protected by the 'cone of protection' effect and between smaller buildings shielded cables up to 42.7 m (140 ft) with each end of the shield bonded to the building's grounding electrode will be adequately protected.

Aerial cables

Aerial cables have protectors fitted that prevent the cable voltage or current rising above preset limits – a sort of fuse but we'll look at these soon. Telecom cables are also fitted underneath power cables so, hopefully, they will take the strike instead of the telecom cable.

Grounding of power systems

It is important that all electrical grounds, metal cold water pipes, the grounding electrodes and structural steelwork is held at the same potential by permanently bonding them altogether before connecting them to

157

a grounding point. This must enable excess currents due to fault conditions, induced voltages and lightning strikes. The bonding conductors (BC) must be able to carry any current likely to be imposed.

Conduit should not be relied on as the sole source of grounding. It is an addition – not a replacement for grounding cables.

Telecommunications grounding and bonding

As with lightning protection systems, the main benefit of telecom bonding is to ensure, as near as possible, that all the ground points are maintained at the same potentials. This is called 'equalization'. The bonding conductor should be grounded at each end wherever possible.

The equipment grounding conductor should be capable of carrying as much current as the conductors carrying the power to each device. This ensures that in the worst case of a short circuit between the mains supply and the ground, the ground wire will be able to carry all the current and will not overheat and fail. We don't want the safety circuit to be the weakest link.

All pieces of equipment in the same system should use the same grounding point and they should be connected in a star configuration. This means that a ground disconnection only affects one piece of equipment, whereas if they were daisy chained (connected one after the other) one disconnection could leave a whole system unsafe.

Each of these grounding systems should be connected back to the main building ground point.

In order the keep any two points at the same voltage, we have to have the lowest possible resistance between them. At moderately low frequencies, the resistance is dependent on only three things: the material used, the length of the connection and the cross-sectional area of the core. So our first choice is always copper and the size is generally 6 AWG (4.1 mm/0.16 inches). At very high frequencies, there is an increasing benefit in using a flat copper strip.

Running the bonding conductor as close as possible to the telecom cable offers a degree of shielding by allowing the bonding conductor to carry some of the short duration interference signals called transients instead of them all appearing on the telecom conductors. The closer the bonding cable is to the telecom cable the better, as it will improve the electromagnetic coupling between the two cables. The effect of this is that any incoming interference will appear equally on each line and will (theoretically) cancel completely at the receiving end – cancellation is never perfect because we never achieve 100% coupling. The mechanics of the situation is that if we have a 1 V positive signal induced onto the telecom signal and, let's say 0.9 V is imposed on the bonding conductor, so at the telecom equipment the input will rise by 1 V but the ground potential will follow it by 0.9 V. In the opinion of the telecom equipment, the input will appear to have increased by only 0.1 V compared with its ground potential.

Grounding choices

We can either use an existing ground point or we can provide our own.

The best option is to use the grounding system that has been installed for the electrical supplies for the building. There may be exposed metal service raceways with an approved bonding connector, an external connection point on the power service panel or a conductor connected to the grounding electrode.

In the unlikely event of additional grounding being required, we cannot use any electrodes or down conductors that are used as part of a lightning protection system nor can we add a new electrode within 1.8 m (6 ft) of an existing electrode. We can no longer use any service pipes as a grounding point. The use of plastic pipes for water supplies means that, although a system may be grounded to a metallic water pipe, it may be only a short distance before we find it attached to a plastic section. Conductive plastics do not have a low enough resistance to be of any use for grounding. They are only used as an anti-static defense.

An example of a grounding conductor is shown in Figure 12.10.

Figure 12.10
A telecom
grounding
electrode

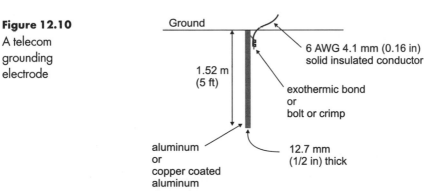

Batteries can be dangerous

Uninterruptible power supplies (UPSs) often use batteries as the back-up energy supply. If they are going to perform this function, they must have enough energy stored to enable them to run all the telecom equipment for several hours. This is a lot of energy and we need to ensure that it is not unleashed accidentally.

Batteries are a way of storing chemical energy that can be converted to electrical energy when required. When different chemicals are brought together in a circuit, they usually generate a voltage between 1 and 3 V, the exact value depends on the chemicals used. Lithium batteries are the current winners in the high-technology stakes, with 3 V outputs and a shelf life of over 20 years!

The single unit that produces voltages from chemicals is called a 'cell'. If we want more voltage, we have to connect cells in series and we then have a group of cells or a 'battery of cells' which we now abbreviate just to 'battery'. So, strictly speaking, when buying one we should not ask for a 1.5 V battery but a 1.5 V cell but this tends to cause more confusion than it is worth (see Figure 12.11).

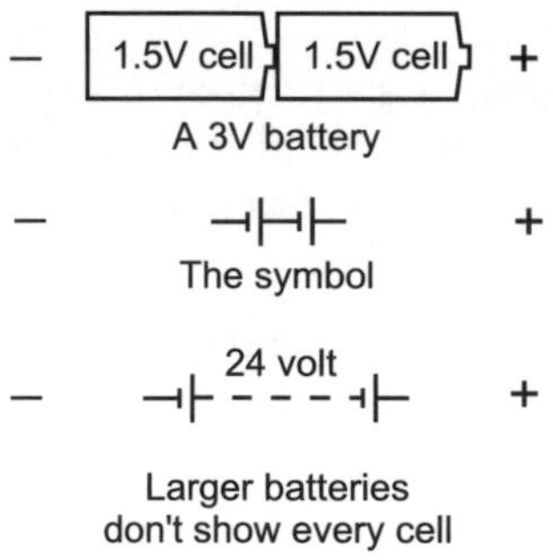

Figure 12.11
A battery of
cells

Batteries are subdivided into primary and secondary. The difference is that only a secondary cell can be recharged. There are two popular secondary cells – lead–acid and nickel–cadmium.

Lead–acid batteries

These contain plates of a spongy form of lead dioxide connected to the positive terminal and plain lead for the negative terminal. When dilute sulfuric acid is added, each cell produces between 1.93 and 2.13 V (we take it as 2 V on average). A bank described as a 120 V bank consists of 60 cells.

As lead–acid batteries are discharged, both plates are coated with lead sulfate which can be converted back to the original material by charging – this is by using an external power supply to reverse the direction of current flow through the battery. If the battery is left for a long period in the discharged state, the lead sulfate becomes impervious and it cannot be converted back to the active materials by charging and, effectively, the battery is dead.

Lead–acid batteries come in two flavors – the first goes under the names of 'wet' or 'flooded'. The other type is variously called 'gel' or 'gelled electrolyte' or 'sealed', even though it is not sealed; 'maintenance free', even though it isn't maintenance free; or finally and technically the best description (though least used) is 'valve-regulated lead–acid' (VRLA).

The first type has the plates submerged in dilute sulfuric acid. The strength or specific gravity is about 1.280 when fully charged but this varies according to the design of the battery and the conditions of use – particularly the temperature. Always check with the supplier.

The maintenance required is to check the level of the electrolyte and top up with distilled water or 'topping-up liquid' as recommended by the supplier. This is to ensure that the plates remain submerged. Wear eye protection – we don't carry any spare eyes. In addition, we should check that the terminals are clean, tight and lightly greased. Maintenance-free batteries do not need topping-up but the other maintenance is just the same.

When the flooded-type charges, they give off gas and a fine suspended mist of sulfuric acid. From the positive plate, we get oxygen and from the negative, hydrogen. This is an explosive mixture and we really don't want anything that contains sulfuric acid to explode, so we need to take precautions. We must ensure adequate ventilation to remove the gases. We must not introduce any form of ignition, which is why the terminals are kept tight (and why we don't smoke near a battery!).

The VRLA or sealed types still produce gas if they are charged at too high a rate. The valve is a pressure release valve that allows the gas to escape. If the valve fails, the pressure will increase until the battery explodes. Not a nice thought.

The other problem with all lead–acid batteries is that they are very heavy so, before installation, it is a wise move to ensure that the floor load will not exceed the designed limit.

Batteries used in UPSs will normally be float-charged, that is, connected permanently to a voltage source so that they are maintained in a fully charged state ready for action if the UPS is called into service. We should check this charging voltage otherwise there could be serious consequences. If discharged, they should be recharged immediately.

The average state of charge of a lead–acid battery makes an enormous difference to its life expectancy, as is shown in Figure 12.12.

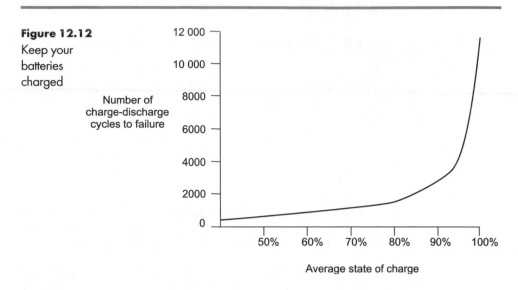

Figure 12.12
Keep your
batteries
charged

Nickel–cadmium cell

This secondary cell, usually abbreviated to nicad or Ni–Cd, has a voltage of only 1.3 V per cell but has a high current capability and is used in some portable rechargeable tools and instruments. Apart from correct disposal, the only hazards are risk of fire or explosion if they are short-circuited or if they are in a fire. Don't carry them in a toolbox where the tools may cause it to short circuit – and worse still, don't even think about putting them in your pocket with a bunch of keys ... The cell is sealed but has a built-in safety valve but if that fails the consequences can be serious.

Alkaline cell

This is a primary cell, available in the same sizes as nicads but they have a terminal voltage of 1.5 V and hold over twice the charge, so the above safety precautions apply with even more emphasis.

Fire precautions

There is no building or any material that is fireproof. The best we can do is to limit the spread of fire from reaching critical areas and provide enough time for evacuation. As we think about fire it is good to remember that more people are killed and injured by smoke and fumes rather than the fire itself.

Each country has its own fire prevention regulations that we must adhere to but the basic ideas are similar.

Containment

To stop a small fire becoming a big one we have to detect it then contain it with fire-resistant materials so it cannot spread.

The building is divided up into fire zones, which are the areas that can contain a fire or protect it from an external fire for a known period of time. The design of the walls, floors and ceiling are fire-rated. This means they are constructed using materials and techniques that can withstand a fire for a given period of time, sufficient for the people to escape or to be safely contained until the emergency services tackle the fire.

There is no point in having a fire-rated wall if we then fit a door that is not fire-rated or a hole is drilled through it to enable a cable or plumbing pipes to be passed through. Unfortunately, we cannot pass cable around a building without making holes through most walls, floors and ceilings. If we avoid it during construction, it is still likely during the life of the building as modifications are made.

Fire stopping

The responsibility of restoring the fire-rating in any barrier lies with the person who drills the hole. Unfortunately, it may happen that the person does not know how to perform this function or for some other reason the hole is overlooked and left open. This can have a major effect on the process of containment, particularly as we remember that smoke and fumes are likely to be a significant threat to the occupants.

We have two types or classes of firestop systems, which are referred to as mechanical or non-mechanical.

Mechanical systems

These are pre-manufactured to fit standard cables, tubes and conduits. The spaces around the cables, etc. are filled with an elastomeric material. This is a soft rubbery material which has high elasticity to fit hard up against any assortment of the standard profiles. They may be fitted within a frame or designed to fit within standard conduits, sleeves or cored holes. Whichever method is used, some means of applying pressure to the elastomeric material is included, as this pressure provides the seal (see Figure 12.13).

Non-mechanical systems

In some ways these methods are easier and better than the mechanical versions. Basically, it uses some pliable material like putty which can be pushed into the area around the cables to block up the holes. This can accommodate irregular-shaped holes.

Figure 12.13
Mechanical
firestops

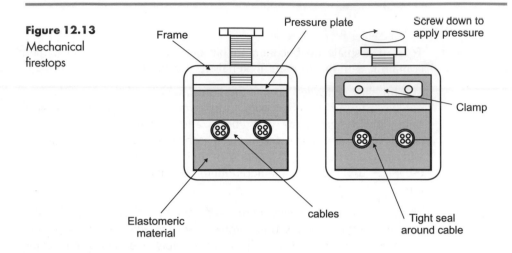

Frame

Pressure plate

Screw down to
apply pressure

Clamp

Elastomeric
material

cables

Tight seal
around cable

Materials that we may meet

Intumescent material

This is useful stuff. Its trick is to expand to several times its volume as
soon as it comes into contact with fire. It comes in strips that we can
wrap around plastic pipes and cable bundles that may burn away to
leave a hole. It's also available as sheets, ready-made flexible blocks,
spray-on foam or even paint. The sheets can be used to block large
holes, particularly when it is backed by a metal sheet.

Cementitious materials

This is very similar but *not* the same as a grouting plaster that can be
obtained as a dry powder or premixed with water and used to fill
large openings. Grouting is likely to crack and fall out when subjected
to fire.

Putty

These are soft, moldable materials that do not set hard but remains
pliable for life. Some types set into an elastomeric solid but both
types can be pushed with hand pressure into small holes around fittings.
Many versions have some intumescent properties and so swell up in fire
for a perfect seal.

Caulk

These are similar to putty but are applied by standard caulk tubes and
adhere to most surfaces, set quickly and do not run, so they can be
applied from the underside of the joint. Some are intumescent, but
not all.

Heavy cable installation

Anything that includes the word 'heavy' should make us think about hard hats and safety boots.

High-pair-count cable runs between floors in slots or sleeves as we saw in Chapter 10. It's nice to think we have a choice of lowering the cable from the top of the building or pulling it up from the bottom but often there is a problem in getting the cable drum up to the top as they are large and heavy, and once up there, difficult to maneuver.

The cables are heavy and must be safely secured and managed during the installation process otherwise there is a serious risk to those involved. The cable may also be damaged because the weight may cause slippage of cable pairs or stretching of the copper.

Prepare the scene by clearing hazards on each floor and placing cones and perimeters to keep people clear of danger. Tie back any existing cables to prevent them becoming damaged and tangled during installation. Make sure all personnel involved are trained and authorized for the operation.

The rolling hitch

This is a useful knot that can be used as an emergency support for the cable. It also enables the cable to slide through in the other direction – a bit like a one-way device for the cable. The knot is shown in Figure 12.14. Lift the knot slightly to allow the cable to slide up and release the knot to lock the cable. We must keep our brain switched on at all times otherwise we run the risk of trapping our hand and receiving serious injury.

Figure 12.14
A useful hitch

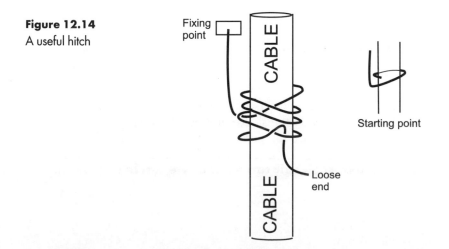

165

Lowering cables

Position the reel away from the slot or sleeve and guide the cable by passing it over a sheave or a shoe. A sheave is a rotating pulley and a shoe is a fixed sliding guide to ensure that the cable will not be damaged by any excessive bending or abrasion on the edge of the slot or sleeve.

Ensure that braking methods are set up, then unreel the cable slowly, guiding the cable from floor to floor. Once it reaches the bottom, fix the cable in position, working up towards the reel. By fixing in this order we can easily pull off more cable if we need it where the cable straps are fixed to a wall which is not immediately next to the slot or sleeve. If we worked downwards, we would need to readjust all the higher ones if extra cable were needed at any point.

Raising the cable

This is more difficult because we are fighting gravity all the way, rather than making use of it. To lift the cable, we will have to fix a power winch at the top of the lift, lower a pulling line and use the manufacturer's pulling eye to attach it to the cable. Finally, set up braking measures as we did when lowering the cable.

Pull the cable slowly and guide it carefully at each floor level as before. As soon as the lift has been completed, secure the cable in position, starting from the bottom as we did before. To provide slack, the cable fixing point should be below the point where the slack is required.

Chapter 12 quiz

1 **If vertical cable requires some slack, the cable straps should:**

   (a) not be used.
   (b) be fixed below the point where the slack is needed.
   (c) be attached to the sleeve and not the wall.
   (d) be fixed above the point where the slack is needed.

2 **The lowest electric current that we can feel is about:**

   (a) 0.1 mA.
   (b) 5 mA.
   (c) 15.5 mA.
   (d) 18 mA.

3   A battery stores:

(a) electricity.
(b) liquid.
(c) UPS.
(d) chemical energy.

4   An intumescent material:

(a) is a powder which, when mixed with water, is edible.
(b) becomes cold when heated.
(c) expands when heated.
(d) is used as a lubricant when pulling cables.

5   The best option for grounding the telecom equipment is to connect it to:

(a) any part of the lightning protection system.
(b) the ground close to, but not touching, any other grounding system.
(c) any piece of metal in the vicinity.
(d) the grounding system installed for the electrical services.

13

A brief introduction to fiber optics

It is quite certain that we are going to come across optic fibers while we are dealing with copper cables, so it is a good idea for us to have a brief look at how they work and what they can be used for – as well as what is likely to damage them.

Here are a few answers to questions about fiber optics.

Why do we use optic fibers?

They have some good points. Because they do not conduct electricity they are immune from electromagnetic interference (EMI) and crosstalk however strong the fields and because the glass is an insulator there is no worry about short circuits. Even the strength members within the cable do not have to be metallic. Compared to copper, they have very wide bandwidths and low losses.

Since they do not radiate any magnetic fields, it is (almost) impossible for someone to eavesdrop on the signals without being detected.

If we compare cables with the same communication capability, the optic fiber is thinner and lighter, which has many knock-on benefits like easier installation and more cables in a duct and lower transport costs.

What is it?

Optic fiber is a transparent material made of plastic or glass that we can use to shine light down. Plastic is cheap, easy to use but has high losses whereas the glass versions are fairly expensive but have extremely low losses – much lower than copper cables. Optic fibers are made from silica glass, which is exceedingly clear and nothing at all like the glass we have in our windows.

How clear is clear?

Amazingly. If the oceans were made of silica glass we would be able to see down to the bottom of the deepest ocean. We are used to our windows being clear and hardly notice the glass but we could make our windows 5 km (or a bit over 3 miles) thick from silica glass and they would still be as clear. That's how clear.

What is the difference between optic fibers and fiber optics?

Optic fiber is the transparent stuff that we can use to transmit light but fiber optics is the system or branch of communication engineering that uses optic fibers. So, we can say that optic fibers are used in fiber optics.

The word 'fiber' is used as a friendly abbreviation for either optic fiber or fiber optics so we can say 'fiber is used in a fiber system'.

How thick are they?

The plastic versions are thick by fiber optic standards and are typically 1 mm (40 thousandths of an inch). In the glass versions they are usually 125 μm thick – this is very thin – about the thickness of two and a half pages of this book. The old term for a micrometer (μm) was a 'micron' and we still see it used from time to time.

Are optic fibers dangerous?

Yes, they probably are. Once we have stripped off the outer jacket and the buffer we are left with a piece of glass only 125 μm thick. This can easily penetrate the skin and break off inside. Once inside it does not come out and is able to migrate around the body. Where it gets to, and what problems it will cause will become apparent in due course but one thing's for sure – it won't do us any good.

What are optic fibers used for?

They are used for two purposes. They can be used to shine light down for illumination. The advantage of this is that the light source can be kept well away from the light output, so if we want to illuminate the inside of a gas-filled container, there is no fear of an explosion. The optic fiber is flexible so the light can be made to go round corners as in Figure 13.1. When used for illumination, we can use either plastic or glass depending on the length of the fiber – glass for longer distances.

Figure 13.1
Light can go round corners

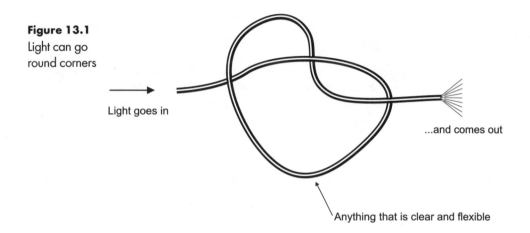

Light goes in

...and comes out

Anything that is clear and flexible

Optic fibers usually have a circular cross-section but if we change the shape we can cause the light to leak out all along the fiber. This obviously decreases the range but it makes the whole length of fiber glow in the dark. When used for illumination, the thickness of the fiber can be increased to a diameter of 1 inch (25 mm) or more. This has many possibilities, we can use it as part of a fire escape to indicate the way out of a building or an aircraft by illuminating the edges of the route. As the light is totally safe it produces no hazards if vapor or water is present. Being safe in water allows us to use optic fibers to illuminate the bottom of swimming pools, which is a significant safety factor.

The second purpose is for telecoms. If we transmit a digital signal along a copper cable, we apply the signal voltages at one end and attach a receiver at the far end. Using an optic fiber, we would use the digital voltage signal to switch an infrared light source on and off, the light would then travel down the fiber and when the flashes of infrared light come out at the far end they will hit a photoelectric cell that changes the light into voltage signals (see Figure 13.2). We use infrared light as this results in fewer losses and hence greater ranges. Infrared covers a wide range in the spectrum and only a part of it is heat – and we don't use that part. Infrared is not visible to our eyes.

Figure 13.2
Copper and
fiber telecoms

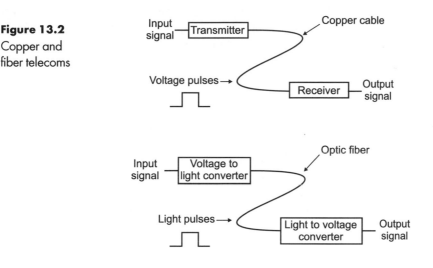

For long-distance and high-speed communications we always use glass fibers as they have very much lower losses than the plastic versions and the light source is always a laser.

What makes the light stay in the fiber?

In Figure 13.3 we are looking at a starfish that is underwater. From the direction that we are looking, the starfish appears to be in position 'A' but in reality this is an illusion because the light is being bent or refracted at the water surface – we are actually seeing light from the real starfish which is at point 'B'. This effect also makes the water seem shallower, as we have all noticed in a swimming pool.

Figure 13.3
Diffraction of
light

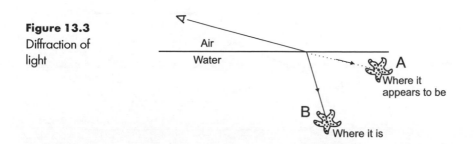

When light moves from a transparent and denser material to a material that is less dense, it is bent just as the light from the starfish was bent at the water surface.

In Figure 13.4, we see a light rays arriving at the boundary at different angles. Notice how we get to an angle where the light cannot cross the boundary. This critical angle is given the rather sensible name of 'critical angle'. Light approaching at an angle beyond the critical angle is reflected back as if from a mirror.

Figure 13.4
Beyond the critical angle, the light cannot escape

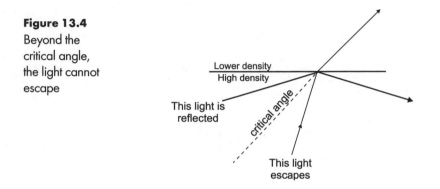

To make an optic fiber, all we have to do is to cover some clear material, glass or plastic, with another layer of glass or plastic, called 'cladding', that has a slightly lower density. Technically, the factor that matters is called the 'refractive index' rather than 'density' but the two are related so it's not that important. If we shine a light into the end of the fiber, any light arriving beyond the critical angle will be reflected back into the fiber. It will then cross the fiber at the same angle and 'bounce' off the other side all the way along the fiber, as shown in Figure 13.5. Any light on the wrong side of the critical angle is not reflected and escapes immediately.

Figure 13.5
The light travels along the core

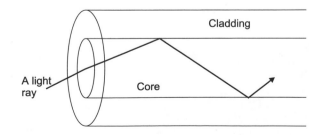

What else do we need?

We have seen that the central fiber passes the light and the outer cladding provides the change in refractive index so that the light cannot escape. In a glass fiber, there is a soft plastic coating around the cladding called the buffer. This is to prevent the cladding getting scratched as it is handled and installed. The slightest scratch will cause the glass to break. All the outer layers beyond the buffer are just like the jackets on a copper cable and may include all the usual things like waterproof layers, rodent proof and metal reinforcing according to the use of the fiber. The layers are shown in Figure 13.6.

Figure 13.6
All the various layers

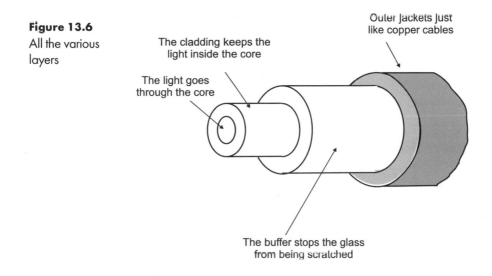

Outer Jackets just like copper cables

The cladding keeps the light inside the core

The light goes through the core

The buffer stops the glass from being scratched

Is the size of the core important?

Yes, particularly so if the fiber is to be used for transmitting telecom signals. One thing that happens in both voltage pulses on copper cables as well as light pulses on fibers is that the square pulse tends to spread out and become rounder as it moves down the cable. This spreading is called 'dispersion' and we can overcome this by installing repeaters at intervals along the route. They detect the pulse and then generate a new re-shaped copy to be passed down the next section of cable. On copper systems, the repeaters can remove any electrical noise from the signal when it regenerates the signal. This is why digital telephone quality is so good on long-distance routes.

In an optic fiber, we can significantly reduce the dispersion by restricting the number of rays of light that travel along the fiber. Rays of light that enter the fiber at different angles are the main cause of dispersion and the diameter of the core is the main limiting factor on the number of rays.

Cores with a diameter of 50 μm or 62.5 μm allow many different rays or 'modes' to travel along the fiber and are referred to as multimode fibers. Reducing the core until it has a diameter of only 8 μm allows only a single mode or ray to travel along the fiber. These singlemode fibers have very low dispersion and hence the data can be sent a much greater distance before we need a regenerator.

The core of a singlemode fiber is roughly one-sixth of the thickness of a page of this book. No wonder that we need a microscope to examine it!

What light source and light detectors do we use?

The usual answer is a laser for singlemode fiber and a light-emitting diode (LED) for multimode fibers. The laser has two advantages over the LED. It has a very narrow beam width so more of the light can be fed into the fiber and the light includes only a very narrow band of frequencies (or wavelengths) and this causes less dispersion.

Are lasers dangerous?

Yes. They can be.

Are all lasers dangerous?

No. It depends on the power and the wavelength being transmitted. Any piece of equipment containing a laser carries the symbol shown in Figure 13.7 which will include the laser classification.

Lasers are classified into groups or classes

When using equipment and finding that it contains a class 1 laser, it is time to relax. A class 1 laser can do no harm providing we use the equipment in the expected manner. Domestic things like CD players contain a class 1 laser.

Class 2 layers give off visible light and it is thought (we aren't totally sure) that the light is so bright that we would be protected by the 'blink' reflex – we have a quarter of a second to close our eyes.

Class 3 give off invisible light and may be hazardous but does not pose a fire risk. We should not look at the laser or the fiber nor view it through instruments. It seems strange and unfair that we can permanently damage our eyes with 'light' which is invisible. As the light is not visible we must be absolutely certain that an optic fiber is not 'live' before looking at the end of it. Without a live laser detector there is

Figure 13.7
Laser warning
labels

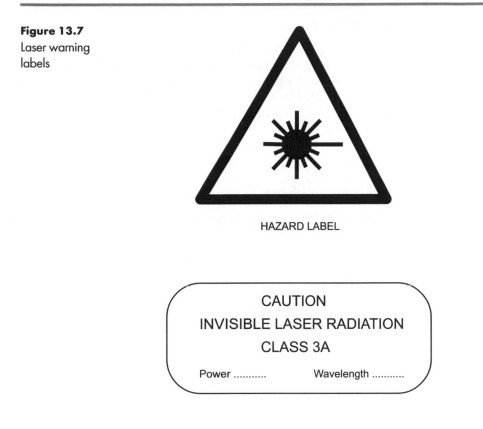

HAZARD LABEL

CAUTION
INVISIBLE LASER RADIATION
CLASS 3A

Power .... Wavelength ....

no way to tell whether a fiber is powered. Make sure that the detector is working.

Never, ever, believe anyone who assures you that the fiber is safe. Check yourself – they may have tested a different fiber, they may have made a mistake, they may be incompetent. Remember that it's your eyes that are at risk and laser damage is not reversible. A letter of apology is no substitute for an eye.

Class 4 lasers are hideously dangerous to eyes and skin and they are a serious fire hazard. Even reflections from surrounding surfaces are dangerous. They should only be used with a system of interlocks to make it physically impossible to be near it without the power being off. Leave it to a specialist. Just writing about them makes me worried.

How do we recognize a fiber optic cable?

It will have 'Fiber Optic Cable' printed along the outside of the jacket otherwise, without opening it up, there is no way to be sure.

Will it break if I bend it?

Yes, if we make a bend too tight. As we bend fiber the first result is that the signal losses increase and, finally, the fiber breaks. The limits are similar to those of a copper cable. As a rough guide, multiply the overall diameter of the cable by 10 and this gives the minimum bending radius but we should always check with the manufacturer's data. A full-sized example is shown in Figure 13.8.

How can I find out more about fiber optics?

A small advert: there is our sister book by the same author and publisher called *Introduction to Fiber Optics* (2001).

And finally ...

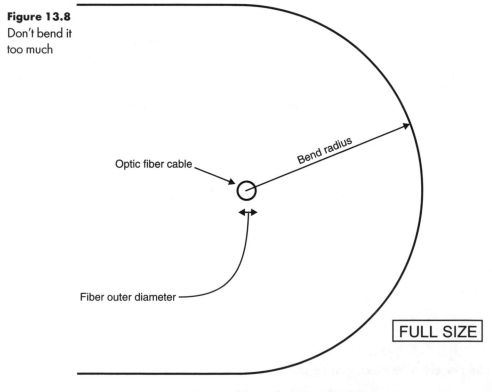

Figure 13.8
Don't bend it too much

Optic fiber cable

Bend radius

Fiber outer diameter

FULL SIZE

Always follow manufacturer's advice

Wrong things that people say about fiber:

1. 'The light goes down a hole in the center of the fiber.' It sometimes looks like it from the diagrams but optic fiber is solid, the light goes down the solid core and reflects off the change in refractive index between the core and the cladding.
2. 'The core must be a good heat conductor.' This has probably been reasoned out because infrared heaters make us warm or cook our dinners. The infrared light that we use is not heat, so there is no heat involved. Incidentally, the glass used is a good heat insulator.
3. 'Glass is waterproof so we don't need to enclose it in a plastic jacket.' Glass is not waterproof. Water causes loss of signal strength so we still have to enclose the cable in a waterproof jacket just like copper.
4. 'If you try bending glass it is obvious that it will break, so optic fibers must be kept nearly straight.' We have all noticed that many plastics are very strong unless we cut or tear a small nick in it to give it a start and then it tears very easily. Much the same occurs with glass. If the surface has no scratches, glass is strong and flexible and will easily bend in a complete circle but the smallest microscopic scratch is enough to destroy the strength just like a tiny cut in many plastics. The real problem is that microscopic scratches are enough to destroy its strength, so the buffer must be put on the glass before anything touches it.

Chapter 13 quiz

1   The buffer:

(a) can be removed once the cable has been installed.
(b) is used to transmit the infrared light.
(c) is used to protect the optic fiber from mechanical damage.
(d) protects the fiber from EMI.

2   The most dangerous laser is referred to as:

(a) class 1.
(b) class 2.
(c) class 3.
(d) class 4.

3   A fiber optic communication system is most likely to include:

(a) a light source, a coaxial cable and a jacket.
(b) an optic fiber, a laser and a photoelectric cell.
(c) an LED, a core and a light source.
(d) a microphone, an optic fiber and a light detector.

4   A singlemode fiber used for long-distance telecommunications is most like to have a core diameter of:

(a) 8 μm.
(b) 50 μm.
(c) 62.5 μm.
(d) 100 μm.

5   Starting from the outside, the parts of an optic fiber are:

(a) core, cladding, buffer, jacket.
(b) buffer, jacket, cladding, core.
(c) jacket, buffer, cladding, core.
(d) cladding, jacket, buffer, core.

14

Moving on

Having got this far, we have made a good start with copper cables and we have a foundation to build on, which we can do with little or no effort.

Just like advertisements, if we are exposed to something for long enough it will be absorbed whether we make an effort to learn or not. All we have to do is to make sure we are exposed to it as often as possible. What we need is a constant stream of information, which we can generate by saying 'yes' every time there is a chance.

Some things are obvious, like books. A good idea with books is to read them through without worrying too much about the bits that don't seem to make sense, which may be the majority if it is a new subject. Don't worry. Just leave it for a month or two, then revisit it for another go. This time a much larger proportion of it will make sense.

Once the basics are sorted we can go a bit further with a slightly more advanced book and do the same with that one – have a skim through, picking out bits of interest and leave it to settle for a while, then have another go at it.

There are other places we can go to gather information, as we can see below.

Exhibitions

If there are any exhibitions advertised, try to go to them. Be warned though that there will be whole chunks of it that you won't understand. Don't worry – they won't make us sit an exam on the way out.

Have a nice time wandering around with a companion, looking at anything of interest, watching demonstrations and picking up leaflets and catalogs. Take it all home and read every word of it, whether it seems to make sense or not. Later, some of the information will fit in with something that we have seen or read elsewhere and it will start to fit together. Adverts are a very useful source of 'guaranteed up-to-date' information.

Catalogs

Catalogs are invaluable. They contain lots of information about the products and their specifications. They should be dipped into whenever a few seconds are going spare. Just like the books, it is a good idea to come back to them a month or so later for another look.

Magazines

Read all magazines in your area of interest. They have explanatory articles and more adverts and free catalogs to send off for. Remember that nearly all companies have web sites and some of them are real little gold-mines full of useful information.

Training courses

Take any opportunity to attend available short training courses. We not only get to see the gear that we recognize from the catalogs but we can get valuable hands-on experience. A valuable part of any course is the breaks when we get a chance to talk to the other people. We gather more experience by listening to what they are doing and hearing about their experiences.

BICSI – a fine organization

Copper cables and telecommunications generally are progressing at their usual breakneck speed as indeed is the number of committees and organizations that regulate, suggest and arrange the way we do things.

With the enormous growth in communications and travel, the separate countries of the world are becoming closer and more interdependent. With the Internet it doesn't matter very much whether we send an e-mail within our own country or to the other side of the world, so technical information is becoming more and more harmonized.

Companies are becoming increasingly global. Rather than buying their material from another country, we very often find that they buy or set up a subsidiary company that can guarantee the supplies. This creates some pressure for uniformity in the working practices, telecoms provision and ultimately the installation on both sites.

There is a continuing convergence in the regulations governing telecoms installations.

So, what is BICSI?

It is an association called the 'Building Industry Consulting Service International' that has been established since 1974. Thankfully, it has since changed its name to BICSI.

It has spread rapidly from its US origins to 85 countries and has more than 20 000 members.

What doesn't BICSI do?

It doesn't make a profit. It is not merely designed as a way of converting your income into their income. As no profit incentive is lurking in the background, we can be more confident about their information and advice.

So, what does it do?

It is a telecommunications association, which means that it encompasses the whole range of design, installation and maintenance of both fiber and copper systems. This is important since installations inevitably involve both media.

BICSI collects together good design, working practices, and solutions to technical problems.

With 85 countries represented, even the most xenophobic person must admit that there is a good chance that an engineer in another country may well have solved the problem that is bugging them at the moment.

Why is it good for its members?

As well as providing an efficient way of spreading the good ideas and warning of the dangers of the others, it provides a range of qualifications that are meaningful.

Quite often when we seek professional advice, we find the walls are covered with certificates promising professional qualifications. The more ornate ones are often dated many years ago – that, at first glance, may instill confidence in the number of years of experience the person has. It can also be proof of how long it has been since the person received any up-to-date training.

I have an impressive certificate promising my competence in installing and maintaining radar installations. All it actually proves is that many years ago I was competent, actually pretty good, at radar technology but that was many years ago, and although I haven't touched a radar since, I am still qualified.

On the other hand, BICSI qualifications are valid only for two or three years and can only be renewed after proving that the holders are keeping abreast with developments and are moving ahead with their knowledge. The requirements are very specific and cannot be bypassed by general statements like 'I have read several magazines in the last three years' or 'I am still employed in the industry but, actually, I now work in the accounts office'.

A BICSI qualification is impressive when job-hunting in this field and you can take the first step on the BICSI ladder with little or no experience in the industry. The examinations are based on knowledge contained in their manuals, which are very comprehensive, well produced and easily read – a rare combination.

BICSI also ensures that all members are kept up-to-date with new developments.

Why is it good for customers?

They can be certain that they get just what they expect. Whether it be an apprentice, an installer or a designer of a complete system, they can be sure that the person who turns up to start will actually be able to do the job. We all hope for this when we employ a professional but BICSI customers know it – an important difference.

BICSI specifies good design and working methods that provide a uniform and proven way of providing a telecommunication system. By installing to a known standard, this reduces time, cost and mistakes. Just as important are the problems avoided when the system is upgraded or extended and we have to trust that the original installation was a competent job.

Here are BICSI's contact details:

BICSI World Headquarters, 8610 Hidden River Parkway, Tampa, FL 33637-1000, USA; e-mail: bicsi@bicsi.org.

BICSI Europe Ltd, Dugard House, Peartree Road, Stanway, Colchester CO3 5UL, UK; web site: http://www.bicsi.org, e-mail: bicsi-europe@-bicsi.org.

Bibliography

BICSI (2000) *Telecommunications Distribution Methods Manual*, 9th edn. ISBN 1 928886 04 3.

Crisp, J. C. (2001) *Introduction to Fiber Optics*. 2nd ed., Butterworth-Heinemann. ISBN 0 7506 5030 3.

Groth, D. and McBee, J. (2000) *Cabling: The Complete Guide to Network Wiring*. SYBEX. ISBN 0 7821 2645 6. (A good read – once you finish this book.)

Sinclair, I. R. and Lewis, G. E. (1996) *Digital Techniques and Microprocessor Systems*, Vol. 3. Avebury. ISBN 0 291 39834 0.

Standage, T. (1998) *The Victorian Internet*. 0 75380 703 3. (A good read.)

Web sites

Adverc Battery Management – www.adverc.co.uk

Chronology of Communication Events: Part 1, Optical Telegraphy – http://www.deas.harvard.edu

Mayflex – www.mayflex.com

Technical Surveillance Counter Measures – www.tscm.com (if you are worried about bugs in your bed, this is a good start)

TectoWeld – www.atitec.com

The Siemon Company – www.siemon.com

Glossary

Acceptance test A test to check that the installed or repaired system meets the requirements agreed with the customer.

Access control mechanism The method used to accept or deny access of a device to a LAN.

Access protocol The procedures that enable access to a network such as token passing and CSMA/CD (carrier sense multiple access with collision detection).

Active circuit The information channel that is in use. It refers to channels carrying data/voice/video or whatever.

Adapter card A circuit card installed to provide the physical connection with a network. Also called a network interface card (NIC).

Address An identification assigned to a network device used to identify the source and destination of a message.

Aerial-buried plant A cable run that consists of both aerial and direct-buried cable.

Aerial cable A cable supported above the ground usually on poles, buildings or similar structures.

Air bottle A source of compressed air that can be used to blow an object to pull a drawstring through a conduit.

Air terminal A lightning conductor or lightning rod.

Alternating current A current that is continually varying in amplitude and regularly reverses in direction. The number of full cycles that occur in a second is the frequency and is measured in Hertz (Hz).

Analog A signal or voltage that can vary continuously and can assume any intermediate value. See Digital.

ANSI American National Standards Institute. An organization developing standards for the telecom industry.

Apparatus A single piece of equipment designed for use by the final user.

Application specific cable Cabling designed or installed to meet the requirements of a specified transmission system.

Architectural assemblies Walls, partitions, dividers, etc., that are not load bearing.

Architectural structures Walls, partitions, dividers, etc., that are load bearing.

Armoring Additional protection for cables against physical damage, rodents or severe outside conditions. Usually plastic coated steel and may be corrugated to provide additional flexibility.

Asymmetric digital subscriber line (ASDL) A digital system that provides a high speed link to the end user or 1.544 Mb/s and speeds up to 128 kb/s in the reverse direction.

Asynchronous transmission A digital transmission system in which the data is sent as a series of bits and each group of bits is separated by a start bit and one or two stop bits. See synchronous transmission.

Attenuation A reduction in signal strength measured as a power loss or a voltage loss. Normally measured in decibels (dB) at a specific frequency.

Attenuation-to-crosstalk ratio (ACR) Measured in decibels, it is the difference between attenuation and crosstalk at a particular frequency. This test is used to ensure that, at the far end of the cable, the received signal is stronger than the crosstalk from other cables.

Autotest A facility offered by field test instruments that can run all the required tests on a cable without the assistance (or interference) of an operator.

Avalanche photodiode The form of photocell used to detect light at the far end of an optic cable.

Backbone The cable and pathway that carries the backbone cabling.

Backbone cabling Cables that connect the telecom or equipment rooms and the entrance facility where the telecom cable enters the building.

Backup path A spare channel for data flow that can be used if the first one fails.

Balanced cable A twisted pair cable that carries voltages of equal amplitude but opposite polarities.

Balun An impedance matching device that is used to connect balanced to unbalanced cables such as twisted pair to coax.

Bandwidth A measure of the information carrying capability of a cable or device. It is measured in Hertz and is the difference between the maximum and minimum frequencies that can be passed.

Baseband transmission This is a transmission in which the whole of the bandwidth is used to transmit just one signal.

Baud A measure of the speed of a transmission. It is often mistaken for the unit bits/second because, in some forms of data transmission, they are numerically equal so the transmission rate of 100 baud indicates that the voltage level being transmitted is changing at a rate of 100 level changes per second. Whether this is equal to 100 bits/sec depends on the system of transmission that is being used.

Bend radius This is the tightest bend that can be applied to a cable without a loss of signal strength or physical damage to the cable.

Binary A signal consisting of two different states such as current/no current, voltage/no voltage or any other two states.

Binder group A group of twenty-five wires in a cable with at least fifty pairs in all. Colored plastic is used to distinguish each group.

Block A connecting device to connect one group of wires to another.

Bonding Permanently connecting metallic parts together to form a conductive circuit.

Bonding conductor (BC) A conductor designed specifically for bonding.

Bonding conductor for telecommunications (BCT) A conductor that connects a build power ground to the telecom ground.

Braid Interwoven metallic strands used to provide shielding from EMI as well as providing more flexibility to the cable.

Bridged tap Multiple appearances of the same cable pair at several distribution points.

Broadband transmission Transmission of several or many signals on a single route. All the transmissions share the total available bandwidth (see also Baseband transmission).

Building distributor (BD) The place where the building backbone cable connects with the campus backbone cable.

Building entrance The room or space where the telecom cables enter a building.

Building grounding electrode system A network of ground points for a building, such as metallic pipes, metal building frames, rod electrodes, etc.

Bundled cable Two or more cables bound together to form a single unit before installation (see also Hybrid cable).

Burn-in The time taken for electronic circuits to settle down to their normal operating temperature.

Byte A group or eight bits.

c Symbol for the speed of light 3×10^8 m/s (186 000 miles/s).

C Symbol for capacitance or temperature Celsius.

Cable One or more conductors enclosed by a single sheath.

Cable assembly A cable that has been prepared with connectors at each end (see Jumper).

Cable reel A reel that cable is wrapped around.

Cable sheath An outer covering enclosing the conductor assembly that may include metallic members or jackets.

Cable support Any hardware that is used to support cables such as cable trays, tie wraps, conduits and support hooks.

Cable tray A support with a base and sides that can hold and route a cable.

Calibration Checking test equipment against known input values to vary their accuracy.

Campus A group of buildings that form a single entity such as an industrial park, a university or a military base.

Campus backbone cable The cables that connect the individual buildings on a campus.

Campus distributor (CD) (main cross-connect) The connection point at the start of the campus backbone cable from which the cable connects to each of the buildings.

Capacitance The ability of charged conductors to store an electric charge. Any two or more conductors that are at different potentials store a charge. The amount of capacitance depends on the distance between the conductors, the material between the conductors and their size. Capacitance is measured in Farads (F).

Categories of cables North American and International standards that specify the physical and electrical properties for cables suitable for stated communication tasks. Produced by ANSI/TIA/EIA organizations (see Classes).

Ceiling distribution system A distribution system that makes use of the space between a suspended or false ceiling and the structural surface above.

Cellular floor A floor distribution system in which the cables run in metal or concrete floor cells as a readymade raceway.

Cellular floor raceway Hollow conduits forming part of a floor systematically placed for easy distribution of cables.

Cementitious A powder which, when mixed with water, produces a lightweight mortar that can be troweled to block and smooth areas often as a firebreak.

Central member A strength member in the center of a cable.

Characteristic impedance The ratio of voltage/current flow in a long (theoretically infinite) transmission line. A short length of cable can be made to appear infinitely long by adding a termination impedance equal to the value of the characteristic impedance.

Chemical electrodes See Chemical ground rods.

Chemical ground rods Copper tubes fill with chemicals to leach into the soil to improve the efficiency of grounding.

Classes Application classes for cabling as defined by ISO/IEC 11801, similar but not identical to the ANSI/TIA/EIA categories.

Coaxial cable An unbalanced communication cable with a central electrode surrounded by an insulator and a mesh conductor for screening and outer layers for protection.

Collision A normal event on an Ethernet network where two or more devices attempt to use the bus at the same time.

Conduit A tube through which cable can be pulled.

Conduit run A system of interconnected lengths of conduit.

Cone A safety marker to indicate dangerous parts of an installation area.

Connecting hardware A device or devices used to connect cables.

Consolidation point A place where horizontal cables may be interconnected.

Container A place where telecom cables are terminated. This may be a room, wall panels a cabinet or similar.

Continuity test A test to confirm that a circuit or connection does not include a disconnection or high attenuation.

Crimp Making a connection by clamping a connector to a cable. No solder, heat or glue is employed.

Daisy chained Connecting devices one after the other, in series.

Data Electrical signals that convey information.

Data network An interconnection of cables and equipment that carries data.

Data transfer rate The rate at which data can be sent over a communication system.

DB See Decibel.

DBm Decibels used to represent a power level compared with a standard level of 1 mW. For example, 3 dBm is a power level of 2 mW.

Decibel (dB) A logarithmic ratio of two power levels. It compares the power out of a device with the power in. For example +3 dB means a doubling of the power level and −3 dB indicates a halving of the power level.

De facto standard A design or practice that, by general acceptance, is satisfactory although it is not laid down in any regulation or code of practice.

Degradation A decline in performance.

Demarcation point (DP) The boundary position where ownership or responsibility changes.

Dielectric The insulation between or around conductors such as the insulation around the core of a coaxial cable.

Dielectric constant The amount by which capacitance increases when an air-filled space or a vacuum is increased when the space is filled with a different insulator.

Digital A signal that changes by discrete steps to represent information (see Analog).

Direct-buried cable A telecoms cable designed to be buried in direct contact with the earth (see Underground cable).

Direct current (dc) An electric current that flows in a constant direction.

Dispersion The spreading out of signal pulses as they pass along a transmission line.

Distribution cell A raceway in or just below the floor to route wires or cables to a specific floor area.

Distribution duct A raceway in or just below the floor to route wires or cables to a specific floor area.

Distribution frame A structure with terminations to make interconnections conveniently made.

Drag line A pull cord or line installed in a cable pathway used for pulling cable or rope to pull heavier cable.

Drain wire A non-insulated conductor placed in electrical contact with the cable shield to bond the cable shield to the ground.

Drop ceiling A false ceiling suspended below the structural ceiling to create a space that may be used for air conditioning or for enclosing cables.

Drywall An interior, non-structural wall made from plasterboard.

Dual foil An additional layer of foil around foiled twisted-pair cable.

Duct A single enclosed pathway used for conductors or cables or sometimes air flow.

Duct plug A fitting on the end of a duct or to surround cables within the duct to provide a liquid or gas-tight seal.

Earth current Any current flowing through the earth.

Earth ground An electrical connection to the earth as in 'grounding'.

Effective ground A connection to the ground of a low enough resistance to prevent the build-up of any hazardous voltages in telecom cables.

Elastomeric firestop A flexible material rather like rubber that can be used to block holes or gaps through which fire or smoke may propagate.

Electrical closet A floor-serving facility to contain electrical equipment, controls or connections.

Electrical noise Unwanted voltages or currents that may occur in any electrical circuit or device and may interfere with the wanted signals.

Electrical resistance The ratio of applied voltage measured in volts to the resulting current flow in amps that occurs in any electrical circuit.

Electromagnetic compatibility (EMC) The ability of a device or system to function correctly within a magnetic field and to avoid creating a magnetic field that could cause problems for other devices.

Electromagnetic field The radiation containing both an electric field and a magnetic field such as we have in a radio signal.

Electromagnetic induction Voltages that are caused by a changing or moving magnetic field. This is the cause of crosstalk.

Electromagnetic interference (EMI) Any electrical or magnetic interference that can induce unwanted signals in any surrounding conductor.

Electrostatic discharge (ESD) The discharge of a static electric charge caused by the interaction of two different materials. Lightning is a prime example.

Electrostatic induction Currents that are caused to flow in any conductor that is the result of a changing electric field.

Endothermic Something which can absorb heat, usually as a result of a chemical reaction.

Entrance facility This is usually a room that is used to allow entry for external cables and associated equipment.

Equal level far-end crosstalk (ELFEXT) A normalized measure of the unwanted coupling between a transmitter at the near end of a cable and measured at the far end of an associated cable.

Equipment cable A cable that connects to a piece of active equipment.

Equipment closet A room that houses telecoms equipment and where horizontal cables connect with the backbone cable.

Ethernet A LAN protocol using a logical bus structure and carrier sense and multiple access with collision detection.

Exothermic reaction A chemical reaction that generates heat as in exothermic bonding or welding.

False ceiling A ceiling suspended below the structural ceiling to create a space that may be used for air conditioning or for enclosing cables.

Fanned Separated cable conductors, strands or pairs.

Far-end crosstalk (FEXT) Crosstalk measured at the opposite end to that of the signal transmission.

Field wiring An electrical intended to be made at the time of installation.

Fireproof Material that will not support combustion. Nothing is entirely fireproof.

Fire rating The time in hours that a material or structure can withstand flames and transmission of heat under standard test conditions.

Fire resistance The time in hours that a material or structure can withstand flames and transmission of heat under standard test conditions.

Fire retardant A substance that can delay the start of a fire or slow down the rate of advance.

Firestop A material, device or assembly of parts that can prevent or slow down the spread of fire, flame, smoke or water in a horizontal or vertical direction.

Firewall Fire: a wall that helps to prevent the spread of fire from one fire zone to another; Computers: a program that prevents unauthorized access to stored data.

Fire zone A part of a building completely enclosed by fire rated walls, floors and ceilings.

Fishtape A device or tool that can help to install a pull line along a pathway or to pull the cable.

Flexible conduit A type of conduit which is flexible enough to aid alignment of straight lengths of conduit.

Float current The current that is passed through a fully charged battery to maintain the state of charge.

Float voltage The voltage that is sustained across a fully charge battery to maintain the charge.

Floor distributor (FD) The distributor used to connect horizontal cabling and equipment (see Horizontal cross-connect).

Foil shield A thin metallic tape wrapped around the cable core to act as a shield against electromagnetic interference.

Foiled twisted-pair cable (FTP) A cable with four pairs of insulated copper cable surrounded by a single overall aluminum foil and an outer plastic sheath.

Foiled twisted quad cable (FTQ) A cable with two groups of four conductors, each twisted insulated copper surrounded by an overall aluminum foil.

Frequency (f) The number of complete cycles per second occurring in a periodic wave. The unit is Hertz (Hz).

Fuse An overcurrent protective device that contains a wire that over-heats and finally melts and disconnects the supply (see Fuse rating).

Fuse cable A length of cable that is two gauges smaller than the rest of the circuit to act as a weak link in the event of a serious current overload.

Fuse rating The current rating marked on the fuse, which is the highest operating current for that fuse. For example, a 10 amp fuse can carry 10 amps without a problem but will disconnect the fusing current if its value is higher than that of the fuse holder.

Gateway The interconnection between two networks that are using different protocols.

Giga A multiple equal to a billion (10^9). For example, 5 GHz is 5 000 000 000 Hz.

Grommet A protective edging placed around a hole.

Ground A point of permanent near-zero potential. An accidental or intentional connection between an electric circuit and the earth.

Ground electrode A conductor or group of conductors that provides a connection to the earth.

Ground potential The zero reference voltage used as a base to measure voltages in a circuit. Whether it is actually zero is not important provid-ing that it is the same voltage throughout the circuit being measured.

Ground potential rise The increase in the local ground potential caused by a lightning strike or a serious power fault.

Hertz (Hz) The unit of frequency equal to one cycle per second.

Hierarchical star An extension of a star topology using a central hub. It is used in structured backbone cabling in buildings and campus layouts.

High pair-count cable Cables with a large number of cables split into binder groups of twenty-five pairs.

Horizontal cable A cable between the work area and the horizontal cross-connect (floor distributor). The word 'horizontal' does not imply that the cable must be laid in a horizontal position.

Horizontal cross-connect (HC) (Floor distributor) A group of connectors like a patch panel or punch down block that allows backbone cables to be connected to patch cords, jumpers and equipment.

Hybrid cable A fiber optic cable containing two or more different types of fiber.

Insulation displacement connector (IDC) A wire-terminating connection in which the insulation jacket is cut by the connector where the wire is inserted.

Inter Between. As in interbuilding which is between two or more buildings (see Intra).

Intermediate cross-connect The connection point between the backbone cable from the horizontal cross-connect (floor distributor) and the backbone cable that extends to the main cross-connect (campus distributor).

Internetwork The communication that connects two or more networks.

Intra Within. As in intrabuilding, within a building (see Inter).

Intumescent Material that swells when exposed to heat. Used in firestops.

Isolation gap An intentional gap left in a cable shield.

Jack A female telecom connector having six or eight contact positions not all that have to be used.

Jacket Outer layer of a cable.

Joule The unit of energy. The number of Joules consumed is equal to volts × amps × time in seconds or watts × time in seconds.

J-hook A support for horizontal cables shaped like the letter J and attached to building structures.

Jumper An assembly of twisted pairs used to join telecom circuits at a cross-connect. Typically 24 AWG (0.51 mm/0.020 in).

Keyed A mechanism built into a jack or outlet that prevents misalignment or incorrect orientation when mated.

Ladder rack Similar to a cable tray but with the appearance of a section of a ladder used to support cables.

LAN See Local Area Network.

Lashing Using thin steel or plastic strands to attach two or more cables together or to a support stand.

Latency The time taken for a signal to pass through a device or network. Also called travel time.

Legend The listing and identification of symbols used on a set of plans.

Local Area Network A specific group of computers within a limited geographical area which are connected together and share and exchange data.

Logical topology The actual topology or method used for nodes in a network to communicate. This is not always the same as the way it appears to be connected in the physical topology (see Physical topology).

Low-smoke halogen free flame retardant (LSHF-FR) Cables of this type produce only small amounts of non-toxic smoke. These cables are also flame retardant.

Main building ground electrode The grounding point for all utilities in a building.

Main cross-connect (MC) Used for the interconnection of the entrance cables and the backbone cable (see also Campus distributor).

Main distribution panel The entrance facility for electrical services.

Maintenance hole (MH) An access point to an underground duct system used for access to cables. Also called a manhole.

Manchester encoding An encoding system used for digital transmissions in Ethernet and Token ring LANs. Particularly good at generating timing transitions.

Manhole Alternative name for a maintenance hole.

Media (telecoms) The alternative transmission paths such as wire, cable, radio signals, etc.

Media attachment unit (MAU) A device used to transmit signals between the Ethernet interface and the transmission media.

Megabits per second (Mb/s) A unit used to measure the speed of transmission of data.

Micron (μm) A millionth of a meter usually called a micrometer.

Modem A MOdulator-DEModulator. A device used to convert between analog and digital signals.

Modular furniture The system of low partitions, desks and other furniture used to divide up a large open work area.

Modular jack A female telecom connector having six or eight contact positions not all of which have to be used.

Modular plug A male telecom connector for cable or cord (see Modular jack).

Modulation Methods used to superimpose information signals onto a carrier wave for onward transmission.

Multimeter A test meter, usually handheld, that can measure voltage, current, resistance and usually other electrical properties.

Multiplexing (mux) The combining of several signals so the information can share a common transmission path. At the far end the signals can be separated by a demultiplexer.

Multi-user telecom outlet assembly (MUTOA) A group of telecom outlets arranged in a single assembly housing. Often used in a modular furniture environment.

Mutual capacitance Capacitance that occurs between two points which are part of the same circuit, as with two conductors of a single pair.

Near-end crosstalk (NEXT) Unwanted signal coupling between cables measured at the cable closest to the transmission.

Network A group of three or more devices that can communicate with each other (see Data network).

Network computer Similar to a PC but without local storage facilities.

Network interface card (NIC) A circuit card installed to provide the physical connection with a network. Also called an Adapter card.

Node An addressable point on a network with processing capabilities, such as a PV or a printer.

Noise Unwanted electrical noise on a wire that interferes with the reception or clarity of the wanted signal. May be the result of other transmissions or external noise from atmospheric static as in electric storms.

Nominal velocity of propagation (NVP) The speed of transmission along a cable expressed as a fraction of the speed of light. Typically around 0.7.

Non-return to zero (NRZ) An encoding signal where there are only two voltage levels to represent the two binary levels.

Nyquist criterion The frequency of samples must be at least twice the frequency of the highest frequency component contained in the signal.

Ohm The standard unit of electrical resistance or opposition to current flow in a circuit or material. It is the resistance of a circuit that would allow a current of one amp to flow when a voltage of one volt is applied (see Electrical resistance).

Ohm's law The voltage in any circuit is equal to the product of the current measured in amps and the resistance measured in ohms.

Ohm-meter An instrument that can measure the resistance of a circuit or device. Usually incorporated in a multimeter.

Open or open circuit A discontinuity, whether deliberate or not, in an electrical conductor or circuit.

Open office cabling The cabling that extends from the telecoms room to the open office area.

Open wire Non-insulated copper or copper and steel wire used in aerial connections.

Outer protection An outer protective layer applied on the outside of the sheath of a cable, usually armored wire or metallic tape.

Outlet box for telecoms A metallic or non-metallic box within a wall, floor or ceiling used to house telecom connectors, outlets or transition devices.

Outside plant Telecom infrastructure designed for installation outside of buildings.

Packet A group of bits with routing information that is transmitted and switched as a single entity.

Packet filtering A security mechanism that examines the contents of a packet entering or being used on a network.

Pair Two insulated wires twisted around each other.

Pair count The number of pairs or wire in a cable or the pair identification serving a location.

Pair twist The uniform twist of an insulated pair that serves to reduce the effects of electromagnetic induction.

Paper insulation The use of paper ribbon wrapped in a spiral or laid longitudinally along a cable to provide insulation.

Patch cord A length of cable fitted with a connector on at least one end and used to join telecom circuits at the cross-connect.

Penetration An opening made in a fire-rated barrier.

Penetration seal A firestop system designed to reseal any penetration made.

Physical topology The physical layout of a network defined by the pattern of connections, e.g. star, bus, etc.

Picofarad A sub multiple of a Farad that is the unit of capacitance. A Picofarad is 1×10^{-12} Farads.

Pigtail A short length of cable with a pre-installed connector at one end.

Pinout A wiring diagram showing the correct connections for a plug or jack.

Plastic insulated conductor (PIC) A conductor insulated by a plastic layer.

Plenum An area such as a suspended ceiling that is used for the circulation of air within a building. Cables used in this area must be approved for this purpose.

Point to point Direct communication between two specified points, by cable or wireless link, etc.

Poke-thru A penetration made in a fire resistant floor to allow cables to the inserted.

Polyethylene (PE) An insulation material with very good moisture resistance.

Polyvinyl chloride (PVC) A tough, flame-retardant insulator, though water resistance is not as high as polyethylene.

Power Measured in watts, it is the rate of using energy equal to one joule per second.

Power arrester A protection device used on power lines to limit the surge voltage caused by a lightning strike.

Power sum equal level far-end crosstalk loss (PSELFEXT) A measurement taken at the far end of a multipair cable of the crosstalk occurring due to coupling from multiple transmitters at the near end. It is a normalized computation.

Power sum near-end crosstalk (PSNEXT) The crosstalk power, assuming that all pairs contribute to the crosstalk level. The measurements are taken at the near-end.

Pre-wiring Installed wiring before walls are finished or wiring installed before it is required in anticipation of future demand.

Propagation The movement of signal along or through a medium.

Propagation delay The time taken for a signal to pass through or along a medium.

Protector unit A device designed to protect against an overvoltage or overcurrent situation.

Pull Installing cable by pulling it through a pathway.

Pull cord/string/wire/rope A line installed in a cable pathway used for pulling cable or wire.

Pulling eye A factory installed device on a length of cable designed as an attachment point to which a pull rope can be attached.

Pulling sheave A pulley with a grooved rim to retain the rope or cable used to guide the cable into the pathway.

Pulp insulation The use of wood pulp as an insulation medium for individual conductors within a cable.

Punch down Terminating copper cable conductors on insulation displacement connection terminals by use of a handheld tool (see Insulation displacement connector).

Quad cable A four-conductor non-twisted pair cable with a red, green, black and yellow conductor.

Raceway An enclosed channel to hold cables.

Radio frequency interference (RFI) Undesired radio frequency signals that interfere with the wanted signals.

Reel brake A brake used to limit the speed at which cable can run off a drum.

Reel dolly A jackstand with heels to assist in moving and paying out cable.

Repeater An internet-working transmitter and receiver used in an analog system for amplifying a signal to overcome losses and to regenerate the signals in a digital system.

Resistance The opposition to current flow in a material or circuit measured in Ohms (see Ohm).

Resistance unbalance The difference in the resistance values of the two conductors that form a cable pair.

Retrofitting Updating an already installed system.

Return to zero (RZ) A digital encoding system in which two voltages are used to represent the digital levels and the voltage returning to a zero rest state during each bit time.

Reversed pair A wiring error in which the pins on a cable pair are reversed at one end of the cable.

Right of way A route over or under land through which telecom facilities can be installed and maintained.

Ring A way of identification of one conductor of a pair. Historically, it was the wire connected to the body or ring of a telephone operators plug (see Tip).

Ring network A network topology where all the nodes are connected in a continuous circle.

Ringing tool A device used to remove cable sheaths.

Riser The space used for cable access between floors in a building.

Router An internet-working device used to direct packets from one network to another.

Screen A continuous aluminum wrap or woven shield around the insulated conductors (see Shield).

Screened foiled twisted pair cable (SFTP) A cable with four twisted pairs of screened cable by an overall braid or foil.

Screened twisted pair cable (ScTP) A cable with one or more twisted pairs screened cable by an overall metallic shield (see Foiled twisted pair).

Segment A part of a network sharing an electrically continuous length of cable.

Self-test A feature in most modern pieces of test gear that allows them to perform a series of internal tests when first switched on.

Server A network device that provides and manages shared resources on a network.

Sheath An outer covering enclosing the conductor assembly that may include metallic members or jackets.

Shield A continuous metallic layer around a conductor or a group of conductors to prevent unwanted coupling between the conductors and external fields.

Shielded twisted-pair cable (STP) A cable containing multiple pairs of twisted cables. Each pair is enclosed in a shield and the whole structure is then covered in an overall shield and an outer insulating jacket.

Short/short circuit A low resistance connection between two conductors which can be intentional or accidental.

Shorting plug A device to provide a short circuit at the end of a pair of conductors for testing purposes.

Signal encoding Converting data into a form suitable for transmission over a given medium.

Signal generator A piece of test equipment that can produce signals at any required frequency for test purposes.

Skin effect At high frequencies the current is increasingly restricted to flowing on the surface of conductors giving the effect of an increased resistance.

Sleeve A short length of conduit lining a hole through a wall or floor to enable cables to pass through.

Slot An opening, usually rectangular, to enable cables to be passed through a wall, floor or ceiling.

Sneak current A foreign current flowing through terminals or equipment that is driven by voltages that are too low to be detected by an overvoltage protector.

Spike A sudden increase of voltage or current in a circuit.

Splice A joining of similar conductors.

Split pair The accidental transposition of two conductors of separate pairs.

SRL (Structural Return Loss) A measure of the amount of reflections that occur from the length of a cable but not from the termination. The higher the figure the better the cable.

Start bit A bit used in asynchronous communications to indicate the start of a new character.

Star topology A topology in which all the cables fan out from a central point.

Star-wired ring A physical star configured as a ring. Also called a collapsed ring.

Stop bit A bit used in asynchronous communications to indicate the end of a character.

Store and forward A transmission technique where messages are stored, then checked for errors and then transmitted to their destination.

STP-A An IBM designed shielded twisted-pair cable with an overall shield capable of transmission at 300 MHz.

Structural Return Loss (SRL) A measure of the amount of reflections that occur from the length of a cable but not from the termination. The higher the figure the better the cable.

Stub-out Conduit installed from a wall outlet to a raceway.

Stub-up Conduit installed from a wall or floor outlet to the ceiling space above.

Surface fitting An outlet box for telecoms at the user work area.

Surge arrestor A device to prevent transient voltage surges from reaching electric equipment.

Suspended ceiling A ceiling suspended below the structural ceiling to create a space that may be used for air conditioning or for enclosing cables. See False ceiling.

Synchronous transmission A digital transmission system in which the data is sent at set intervals of time governed by a clock signal used to synchronizing the sending or receiving equipment and thus not requiring any start or stop bits. See asynchronous transmission.

Telecommunications closet An enclosed space for telecom equipment, cable terminations and cross-connects. It can be used as the junction between backbone cabling and horizontal cabling. See telecommunications room and wiring closet.

Telecommunication main grounding busbar (TMGB) A busbar placed in an accessible location and bonded by the bonding conductor to the building power ground.

Telecommunications room An enclosed space for telecom equipment, cable terminations and cross-connects. It can be used as the junction between backbone cabling and horizontal cabling. See Telecommunications closet.

Terminal A device used to connect wires to each other or a point where information can enter or leave a telecom network.

Terminal block A unit that can terminate cable conductors and is a transition point between cable conductors.

Terminator An impedance matching device placed at the end of a transmission line. It absorbs power and prevents reflections.

Thicknet A 50 Ω coaxial cable used for Ethernet transmission.

Thinnet A thin version of the 50 Ω coaxial cable used for Ethernet transmission.

Throughput The amount of data passing a certain point in a given time.

Tie wrap A plastic or hook and loop strip used for binding cables.

Time domain Reflectometer (TDR) A test instrument that, by sending a pulse down a line and measuring the amplitude of reflections, can determine the length of the line and many fault conditions that have occurred.

Tip A way of identification of one conductor of a pair. Historically, it was the wire connected to the end or tip of a telephone operators plug. See Ring.

Token (ring) A signal code that is passed around a token ring. Possession of the token allows the node to transmit a message. This prevents two or more devices from transmitting at the same time.

Topology The physical or communications paths layout of a network.

Transition point (TP) A location in a horizontal cable where a change of form takes place, for example a flat cable connected to a round cable. The 2002 versions of EN50173 and ISO/IEC 11801 have replaced the TP with a CP in the generic cabling layout.

Transmission media The physical carrier of a data transmission such as copper, optic fiber or wireless.

Transposed pairs When two pairs of conductors are terminated in each other's location.

Twisted pair Two individually insulated conductors physically twisted together to form a balanced pair.

Twisted pair cable A multicore cable in which the pairs of conductors are twisted together to cancel the effects of electrical interference.

Underfloor raceway A pathway built into the floor to carry cables that can pop out in the required floor area.

Underground cable A telecom cable designed to be installed underground in a trough or duct but not in direct contact with the earth. See Direct-buried cable.

Uninterruptible power supply A power supply which has a backup source of power that can be connected without any interruption of power to the equipment. It generally uses battery power for the backup.

Unshielded twisted pair cable (UTP) Cable containing one or more pairs of twisted copper but without any metallic shielding.

Velocity of propagation The speed of transmission along a cable relative to the speed of light. See Nominal velocity of propagation (NVP).

Volt The unit used for electromotive force or potential difference. One volt is able to pass one amp through a resistance of one ohm.

Wand A test device that is able to detect a signal placed on a cable to enable it to be identified.

Wide area network (WAN) A network that extends over extended distances using a variety of media such as cables and satellites.

Wire map tester A test instrument that can carry out all the required acceptance testing required on cables.

Wiring closet A telecoms room.

Wirewrap A method of terminating a conductor by wrapping it around a post.

Work area A building space where the end-users interact with the telecoms equipment.

Work area cable or cord The cable connecting the equipment used to the telecoms outlet.

Zone cabling The cabling which extends from the telecoms room to the open office area.

Zone of protection The area that is protected from a lightning strike.

Quiz answers

Chapter 2

1 (b)
2 (d)
3 (b)
4 (b)
5 (d)

Chapter 3

1 (c)
2 (a)
3 (b)
4 (a)
5 (c)

Chapter 4

Formula: $\text{gain} = 10 \log \left(\dfrac{\text{power}_{\text{out}}}{\text{power}_{\text{in}}} \right) \text{dB}$

1 (c) To convert from a log to a ratio, divide the log value by 10 then use the 10^x button to find the ratio.

In this case, we start with $-20/10 = 2$, now 102 = 0.0101 or a reduction of 100 times.

2. (b) Use the formula above and put in the 30 mW and the 1 mW (see notes on dBm).

3. (c) Remember to add decibels to combine the effects.
4. (a) From the formula above.
5. (d) Start with the 23, divide by 10 = 2.3. Use the 10^x button to find the output/input ratio. This gives 199.5, or nearly 200. This means the output power is 200 times larger than the input, so the input is 5/200 = 25 mW.

Chapter 5

1 (d)
2 (b)
3 (d)
4 (a)
5 (c)

Chapter 7

1 (c)
2 (b)
3 (d)
4 (a)
5 (d)

Chapter 8

1 (c)
2 (a)
3 (c)
4 (a)
5 (d)

Chapter 9

1 (d)
2 (d)
3 (a)
4 (b)
5 (c)

Chapter 10

1 (d)
2 (b)
3 (c)
4 (a)
5 (a)

Chapter 11

1 (c)

Remember that for voltages, decibels use '20' in the formula

$$dB = 20 \log \left(\frac{V_{out}}{V_{in}} \right)$$

$$-10 = 20 \log \left(\frac{V_{out}}{5} \right)$$

Divide both sides by 20

$$-0.5 = \log \left(\frac{V_{out}}{5} \right)$$

Take the antilog to get rid of the 'log'

$$10^{-0.5} = \frac{V_{out}}{5}$$

Now use a calculator to work out a value for $10^{-0.5} = 0.3162$

$$0.3162 = \frac{V_{out}}{5}$$

Now multiply by 5

$$1.58 = V_{out}$$

2 (a)
3 (c)
4 (c)
5 (b)

Chapter 12

1 (b)
2 (a)
3 (d)
4 (c)
5 (d)

Chapter 13

1 (c)
2 (d)
3 (b)
4 (a)
5 (c)

Index

Index

Index